# star
## trails

*50 Favorite Columns*
*from* Sky & Telescope

*To Wendee, my beloved wife*
*To Nanette, our daughter, and to Mark, our son-in-law*
*To Summer and Matthew, our precious grandchildren.*

© 2007 New Track Media LLC
Sky Publishing
90 Sherman Street
Cambridge, MA 02140-3264, USA
SkyandTelescope.com

**Library of Congress Cataloging-in-Publication Data**

Star trails : 50 favorite columns from sky & telescope.
    p. cm.
  Includes index.
  ISBN-13: 978-1-931559-46-1 (pbk.)
  ISBN-10: 1-931559-46-5 (pbk.)
    1. Astronomy--Miscellanea.  2. Astronomy--Observations.
  3. Telescopes.  I. Sky and Telescope.
  QB52.S75 2007
  520--dc22

                              2007000084

Printed in China

# star
# trails

*50 Favorite Columns*
*from* Sky & Telescope

David H. Levy

Sky Publishing
A New Track Media Company
Cambridge, MA

# contents

## places

## things

## objects and events

# Preface

In early 1963, while a patient at the Jewish National Home for Asthmatic Children, I joined the Denver Astronomical Society. I was told that as a member, I would receive the magazine *Sky & Telescope* every month. Being only 14 at the time and far from my home in Montreal, getting mail — *any* mail — was an event to be anticipated, and I counted the days until my first copy of *S&T* arrived. I still have that well-worn March issue with its picture of a homemade telescope on the cover — an auspicious image considering the direction my own telescopic interests would later take me.

## First Attempts

In those days my enthusiasm was far superior to my brain or ability. For instance, as a resident of a communal home I had plenty of opportunity to share the sky with the other boys and girls. So I took Echo, my 3½-inch Skyscope, and mated it with Syncom, a 5-inch reflector. The two telescopes together allowed me to double the number of kids who could peer at the sky during one of my "public" observing sessions. I thought it was neat, so I wrote up the idea for *S&T*.

It was rejected, of course, but the way it was rejected increased my respect for those who wrote and worked for the magazine. I received a letter from *S&T*'s founder, Charles Federer, who informed me that while my article was probably not good enough for publication, it was being forwarded anyway to Bob Cox, their "Gleanings for Amateur Telescope Maker's" editor. Sure enough, some weeks later I received the nicest rejection letter from Bob, who commented that he knew how difficult asthma could be and encouraged me to continue my work and resubmit something else later, particularly with pictures.

Another memory dates to late 1981. I left an astro-club meeting feeling frustrated, and when I got home I decided to relieve my aggravation by writing an article for *S&T* about my collection of telescopes. I knew the story would never be accepted but just writing it was cathartic. Imagine my surprise when, a few weeks later, a letter arrived from Steven James O'Meara (the editor of *S&T*'s amateur-astronomers section) announcing that my article would be published in the April 1982 issue!

## Star Trails

Almost exactly six years later (January 1988 to be exact), Star Trails first appeared in *Sky & Telescope.* Initially conceived as a column about the amateur observing experience — think "On the Road with David Levy" — Star Trails has evolved over the years. For a while it took on an astrono-mer-of-the-month flavor, but lately it has returned to its original goal: to introduce the night sky the David Levy way . . . the way I experience it.

And now, more than 200 columns later, the editors of *S&T* and I felt it was appropriate to assemble 50 Star Trails into a book. This is not a purely best-of compilation, though it does include many of my favorite columns. Instead, the articles I've chosen reveal how amateur astronomy and the people who practice it have changed over the years.

While selecting the columns, I noticed that they divide naturally into five topics: Ideas, People, Places, Things, and Objects and Events. "Ideas" looks at wondrous books like *Cole of Spyglass Mountain,* the marriage of art and astronomy, and the concept of Earth striking back at a comet. The "People" section includes my visit to an astronomy club at a federal prison, a letter to my granddaughter, and a look at two great comet hunt-ers. In "Places," I visit Birr Castle in Ireland, an amateur observatory in El Salvador, and Kitt Peak in Arizona (where I spent the night observ-ing). "Things" includes adventures with telescopes and Schmidt cam-eras, while eclipse voyages, a gravitational lens, and a Venus transit are part of "Objects and Events."

Star Trails is a window into the minds and hearts of the people who have made a difference in how we study and enjoy the stars. This book is also a look at the personal experi-ences that have shaped the course of my nights under the stars. It is my hope that these stories will enrich your own stellar evenings.

Here I am, sans scopes, at the Adirondack Science Camp in 1966.

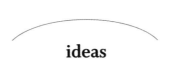

ideas

# Starting Over

Do you ever wish you were starting over again? Many of us begin in astronomy through the mass-production routine of a star party, where we would stand in line at a telescope, waiting for that precious turn at the eyepiece that would whisk us off to a new world. Sadly, that first look turned out to be a disappointment. We wanted to see the sky because we have seen photographs of its wonder. But there, through a tired old instrument with a fingered and filmy eyepiece, we saw a sight that was less than world-class; the magnificent Andromeda galaxy of the 200-inch reflector shrank in size and splendor to a dull, light-polluted glow. If that is our first memory, no wonder a telescope failed to catch us.

The problem with star parties for beginners is that they offer too much too superficially. The night sky becomes a maze and the new observer approaches it like Lord Ronald, a character from novelist Stephen Leacock who "flung himself upon his horse and rode madly off in all directions." It's fine for beginners to try and view everything at once, but only if they are also working on a simple observing project with a clear objective. Such projects offer them depth and meaning — and the potential to expand their interests methodically and deliberately.

There is another way. Have them approach the stars cold, and let the sky be their teacher. Ideally, a new observer shouldn't even have a star chart. What would the sky from, say, Deneb look like to an "experienced observer"? It would be like seeing the sky for the first time, a grand feeling. It is the feeling an Earth bound beginner can still enjoy.

After some 7,500 observing sessions I still observe this way. I don't do my best work the "star-party way," where the telescope is deliberately swung from known object to known object, and the sky politely responds. My favorite sessions are comet hunting where the aim is to discover the unexpected. I begin these sessions with a glance over the entire sky to remind myself what is there. Then, as I uncover my telescope, I block out a large patch of sky, maybe 45° by 10°, where I plan to search. It's a challenging way to observe, because I don't really know what I'll sweep up during the next hour or two. An unfamiliar galaxy may show up, or an open cluster I haven't seen in years. M31 displays added luster when it glides into view, uninvited but always welcome. It is the sky's game, not mine, and every night offers discovery.

Yet, in contrast to this casual approach, I know of amateur and professional astronomers who have gone to extraordinary lengths to make a discovery. It's as if a clear night and a telescope can remove all sense of moderation. Thus, Clyde Tombaugh spent a bitter cold night at the 13-inch astrograph at Lowell Observatory, taking exposures for his search for Planet X. While putting in extra hours trying to keep on schedule, he began to feel strangely comfortable at the guiding eyepiece. When he tried to get up later to change plates, he could hardly move for the cold. "Had I followed the temptation to take a nap during that exposure, I might never have awakened," he admitted at a recent lecture.

The audience of amateur astronomers was captivated by his words, but their individual expressions indicated that risking one's life for an exposure is normal. These amateurs, who would not wish to tell their families about their exploits, freely share their stories with fellow amateurs.

Walter Scott Houston was in the front row, also listening; he has also gone to extremes to make a sighting.

One summer evening in 1957, Scotty was sitting comfortably outside his Manhattan, Kansas, home with his close friend Clifford Simpson, who asked about a bright comet hanging in the western sky. "No, no, " Houston replied, puffing sagely on his pipe and thinking of the just-departed Comet Arend-Roland "the comet's gone." But Simpson persisted.

As Scotty reluctantly turned to chide his friend, he saw a bright, new comet with a long tail. Leaping from his chair, he ripped the canvas cover from his telescope. The comet was setting below a crab apple tree, so Scotty had to work fast. Skillfully he climbed a shaky eight-foot ladder and achieved a precarious balance by holding the instrument with one hand and his pipe with the other. But moving the telescope required both hands, so Scotty put the pipe in his jacket pocket and focused on the comet.

"Scotty, you're on fire!" Simpson yelled. Houston leaned away from the eyepiece, threw his smoldering jacket on the ground, than reached up for another look. Soon the grass around him was ablaze. "We had prairie fires out in Kansas all the time, " reflected Houston, "but never a comet like that!"

Whether you observe comets or quasars my message to you is the same: get out there and let the sky teach you.    JANUARY 1988

### Author Update:

*Nearly 20 years later, the column lives on in* Sky & Telescope *and in the book you are holding.*

# Accidents at the Telescope

Alfred Joy was lucky. He was one of Mount Wilson Observatory's most illustrious astronomers, and he narrowly escaped death in 1946 when he fell more than 20 feet off an observing platform. Even though his assistant got help immediately, there was doubt he would survive his massive injuries. Had he been observing alone, Joy might have died.

Observing can be dangerous. In fact astronomy has been rated the seventh most hazardous white-collar profession. The death of Mark Aaronson in 1987 is a cold reminder. He was crushed by the rotating dome of the 4-meter Mayall reflector at Kitt Peak National Observatory.

Astronomers do take chances whenever they are with a telescope. On Arizona's Mount Lemmon, an observer was busy doing photometry when he heard a knock at the door. It was someone from a neighboring telescope, his face covered with blood. He had walked into a corner of an instrument attached to the telescope. This kind of accident is common, since a heavy instrument bolted to a multi-ton telescope will not budge when a 180-pound observer walks into it. Even I have earned a bruise and a headache that way while imaging Comet Halley through a 61-inch reflector.

Sometimes accidents happen because the observing takes place where it's dark and cold. Long nights under such condition can be taxing on the body and mind. High elevations may be fine for infrared tele-scopes but not for humans used to breathing oxygen in abundance at sea level. Furthermore, the drive to and from mountaintop observatories can be dangerous. In 1986 two people perished when they drove off Hawaii's Mauna Kea after observing Comet Halley from 9,000 feet.

Thanks to the need to point telescopes and their equipment at odd angles, observing platforms and dome walkways can become perilous at any time. Such was the experience of one observer at the Angelo-Austra-lian Telescope. Trying to avoid a jutting instrument, he paraded right off a walkway into the center of a hoist intended to lift the mirror cell. He nar-rowly escaped death. And no, the walkway did not have a rail.

Portable amateur telescopes are a blessing, since they submit some-what if a user bumps into them. In fact I don't consider a telescope re-ally mine until I have bruised myself on it somehow. The worst such incident happened just a few months ago when I was removing an 8-inch Schmidt camera from its fork mount. During this awkward task, the camera slipped and started to fall. By quickly moving my hand between the camera and the base of the fork, I traded one impact for another: camera safe, finger healing.

Comet observer Stephen Edberg of La Canada, California, had an eerie experience once night when a sheriff's truck passed by his roadside-observing site and scanned the lower terrain with a search light. " Later, I heard footsteps down slope," says Edberg, "as if someone was trying to walk unseen." Then the sound slowly faded

The interior of an observatory dome is, by necessity, dark at night. It's also full of odd-shaped equipment (as can be seen here at the Steward Obser-vatory's 61-inch scope), so observers and technicians must be extremely cautious when moving about.

until it was gone. These footsteps were real, but how often have we heard other noises and imagined wild animals or lunatics lurking in the rough, or been startled half to death? Indeed, the perception of trouble can be as frightening as real trouble.

How do we minimize risk? The best way is never to observe alone. If you must go out by yourself, please tell someone where you will be and when you will return. The University of Arizona provides its observers with a special necklace that can alert emergency crews. At least a dome offers some protection from the outside world; a solitary amateur set up in a field is an easy target for attack, even if the telescope in the dark could bear some resemblance to a mortar or other artillery piece!

All other precautions aside, a telescope as a weapons system won't help if Nature itself is the problem. On the evening of September 25, 1982, a mile-long fissure opened on the floor of the Kilauea volcano on Hawaii's Big Island, through which spouted a 200-foot "curtain of fire." As hundreds of people viewed the eruption safely atop the crater wall, Stephen J. O'Meara was strolling the caldera floor with his 10 × 50 binoculars, only 75 feet from the fountains.

"Sitting two feet from a lava fall, I watched the flow of new land pour onto the volcano's floor with pops and gurgles," Steve wrote in his Hawaiian diary. "I decided to walk the perimeter of the flow to the fissure, where the piston activity of the fountains was dying down. At one point I took the opportunity to nova hunt not three feet from an advancing wall of lava.

"Lying supine, glasses raised to the sky, I searched star fields in Monoceros. Despite purple sky, green stars, and white molten rock, I completed the search sheltered from the chilling breeze by the heat of the flow." OCTOBER 1989

### Author Update:

*In June 2006, I casually walked into the railing at the Steward Observatory's 61-inch telescope in the Catalina Mountains. As a reminder of that event (and this particular column), I hobbled about for weeks.*

# The Third Star

It's dusk on a Saturday in ancient Jerusalem. A man stands outside the temple, anxiously looking upward. He sees a bright star rising in the east and a second one overhead. This man knows the sky well, and, as he turns to the northwest, he glimpses Capella in the gathering darkness. "That's it!" he thinks. Three stars have appeared, signaling the end of the Sabbath more than two millenniums ago.

Although the Jewish tradition of sighting stars is no longer generally practiced, it dates back to the dawn of sky watching. (On cloudy nights the observer would look at two strings — one blue, one white — and would judge the Sabbath over when he could no longer tell their colors apart.) This sense of space and time, deeply rooted to the observation of the sky, is one of the hallmarks of Judaism.

For many of us, the meaning of the night sky goes beyond mathematical equations; our interest in the heavens has a strong spiritual component. I realized this many years ago during *Kol Nidre,* a liturgical prayer recited at the beginning of the service on the eve of Yom Kippur. Judaism's holiest day. *Kol Nidre* evening is known for some of the most soaring music of the Jewish liturgy, but for me its meaning extends literally to the sky. While walking home as a youngster after one of these services, I noticed the bright, 10-day-old gibbous Moon dominating the evening sky. I realized that the Moon displays the same

phase every *Kol Nidre* night, as it has through the ages. That moonlit walk home added a strong spiritual dimension to my developing interest in the sky.

A sense of spirituality runs strong in our family, six generations of which have been with the Congregation Shaar Hashomayim in West-mount, Quebec, Canada. The synagogue, whose name means Gates of Heaven, is celebrating its 150th anniversary this year. My grandfather, William Levy, helped design this house of worship, which was built in 1922. The synagogue is especially known for its exquisite choir, whose renderings of Jewish liturgical music have made Sabbath and holiday services a joy to attend. When I'm hunting for comets late at night, I often think of the choir — its pieces end so beautifully and peacefully that they almost command their listeners to gaze heavenward.

Having a spiritual sense of the sky is not just a feeling. In Judaism the relationship is a literal one, since our calendar is based on the orbit of the Moon about the Earth. It's not a coincidence that the Moon is always 10 days old every year after *Kol Nidre* services, nor is it an accident that the total lunar eclipse of April 3rd this year took place on the first Seder or ceremonial dinner of Passover, which always occurs on the night of a full Moon. This spring's eclipse is one of several I have seen during the first night of Passover; in 1968 I rushed away early from a Seder to see one.

Of all the many hours I worked at Palomar Observatory with Eugene and Carolyn Shoemaker, conducting nightly sky patrols with the 18-inch Schmidt, my favorite ones were when the dome shutters slowly opened to reveal a darkening sky. No matter how busy we got during the next 12 hours of astrophotography, I cherished every minute that the opening shutters cajoled the sky to enter. It is a deeply spiritual feeling that is separate from any religious belief, and I'm sure many skywatchers, regardless of their persuasion, have similar experiences at the start of a beautiful starlit night.

Equations can explain the physics of what we see in the sky, but the wonder goes beyond numbers. Each of us has a personal reason for enjoying the precious beauty of the night sky. For some, the backdrop of a liturgy helps. The ancient nomadic Jews depended on the Moon for a calendar to unify the people. Thus the Sabbath and festivals were proclaimed ended after an official observer noticed that the evening sky was dark enough for three stars to appear.

That man who stood outside the temple in Jerusalem, waiting with anticipation for the sky to darken gradually until three stars appeared, must have felt his role in the cosmos. Sabbath didn't end until the sky presented him with three stars. It must have been a singularly personal way to get acquainted with the sky. Seeing that third star must have felt as wonderful as discovering a comet.                    SEPTEMBER 1996

# Cole of Spyglass Mountain

*Star Trails* is not the place to find book reviews, especially if you'd like to learn what is new. This column has reviewed books only twice. In July 1990 it introduced readers to the tragic story of Swithin St. Cleeve, a variable-star observer and the leading character in Thomas Hardy's 1882 novel, *Two on a Tower*. Then in March 1991 it reviewed Leslie Peltier's *Starlight Nights*, first published in 1965, and which I consider to be one of the finest books ever written about the romance and adventure of stargazing. For more about these two books, see pages 102 and 20 respectively.)

*Cole of Spyglass Mountain* is probably harder to obtain than either of these books. It's Arthur Preston Hankins's moving story of the early life of Joshua Cole, and it has been largely out of print since its appearance in 1923. I've known about the novel since I first started looking at the stars in

the summer of 1960. "David," my father said one evening at dinner, "when I was 14 [in 1923] I read the story *Cole of Spyglass Mountain*." Dad went on to describe the story of an impoverished amateur astronomer, also 14 years old, who grew up to make a discovery that was largely ignored by the astronomical community. He recalled the exuberance he felt when, in the book's final paragraphs, the discovery was confirmed. He hoped that I might be able to find a copy of the book so that he could read it again.

Although I never found the book in his lifetime, my friend Peter Jedicke located one in a Montreal library's rare-book collection. Unfortunately, I had enough time during my visit just to photocopy the last page. Finally, in May 1997, Van Robinson, an experienced deep-sky observer I met at the Texas Star Party, was able to find a copy of the book. Thanks to his kindness, I finally have a copy of Father's long-lost book.

The novel reads as if Charles Dickens had made an amateur astronomer out of Oliver Twist. The story opens as Joshua Cole is expelled from school for refusing to let the teacher whip his younger brother. Facing the wrath of his cruel father, who believes in disciplining his son by holding his head under water until he almost drowns, Cole runs away, joining a band of railroad workers. There he meets a young girl named Madge Mundy. One night as the young couple are gazing at the stars a local constable apprehends Cole, charging him with being a runaway. Refusing to deal with his son, Cole's father places him in a boy's reformatory called the House of Refuge, where one loses one's name. Cole becomes known only as 5635. What he doesn't know yet is that his mother's family is very wealthy and he is due a tremendous inheritance when he turns 21.

Beaver Clegg, head of the institution's juvenile department, is an amateur astronomer who belongs to an organization very much like the American Association of Variable Star Observers. During his years at the House of Refuge, Cole becomes an accomplished observer himself, and Clegg eventually gives his telescope to 5635.

Left on his own after his discharge, Cole sets out west to find Madge, whom he suspects might still be working on the railroad tracks with her family. His trip begins well: each evening in a different town, he sets up his telescope by the roadside and shows the stars to passersby for a fee. Then he hitches a ride on a freight train, hoping to make it a little farther west before being thrown off. But one night the telescope, his only remaining hold on the stars and a meager income, is stolen and he continues west alone.

Finally Cole reaches California, and locates Madge and her family working on a railroad tunnel. Near the site is a hill over which is the clearest, darkest sky Cole has ever seen. Working on the tunnel, he is able to save enough money to buy a telescope and set up a homestead on the hill that he names Spyglass Mountain. He has heard about Percival Lowell's theory that a Martian civilization had built a system of canals on the planet, and he resolves to prove it. His plan is to do a systematic study of Mars, looking for patterns that are unmistakably artificial.

Back east, his father continues to prevent Cole from learning about his inheritance. His strategy: to keep his son from returning east by arranging with a neighbor to harass Cole. The neighbor tries to burn down Cole's home and fires shots at his observatory. All the while, Mars is approaching perihelic opposition, the best in years. On the night of June 18, 1922, Cole peers through his telescope and sees nothing unusual. But with improving seeing, he begins to detect the canals. Then, during a brief moment of perfect seeing, the planet shines forth, its canals bearing the unmistakable pattern Cole was looking for. Just at that moment his neighbor's bullets start flying again, and before he is able to confirm his sighting, a bullet strikes him.

Although Cole is badly wounded, he is able to report his observations. At first no one can confirm them. But then the newspaper headlines come in and Cole seems reborn: an observatory "has photographed figure unknown American observer saw on June Eighteenth. . . . Scientific World is Asking Breathlessly: Is Mars a Living Planet?"

My father loved to read, and he never forgot the ending of a good story. Back in 1960 he would sit wistfully at the dinner table, reminiscing about the story of the young amateur astronomer who wouldn't give up. Neither did Dad. As he grew older, he would ask me occasionally if I had ever found the book that had so enthralled him as a teenager. As Father developed Alzheimer's disease, I watched his memory fade away until he didn't seem to remember that I was his son. A few months before he died, however, we were walking together when suddenly he turned toward me. His face brightening, he asked, "Did you ever find *Cole of Spyglass Mountain*? What a story that was!" And then, as if he'd read it yesterday, he recited the ending: "First to report discovery, Cole of Spyglass Mountain famous in a night."

Father never did forget that euphoric tale of a young man and his dream. And thanks to the fact that at least one copy of the book still exists, neither will I.                    APRIL 1998

# A Starlight Night

A starry night. If you're an avid stargazer, as I am, then nothing gets your juices flowing more than the sight of the Sun setting in the west with the promise of a dark, crystal-clear night. And that's precisely how *Starlight Nights: The Adventures of a Star-Gazer* — Leslie C. Peltier's magnum opus — begins. I love the opening lines of the book, and each year, when I lecture to a group of children attending an astronomy camp at the 61-inch Kuiper telescope near Tucson, I begin with those words:

> There is a chill in the autumn air as I walk down the path that leads along the brow of the hill, past the garden and the big lilac, to the clearing just beyond. Already, in the gathering dusk, a few of the stars are turning on their lights. Vega, the brightest one, now is dropping toward the west. Can it be that half a year has gone since I watched her April rising in the east?

The quiet strength of Peltier's words fills the telescope's cavernous dome as I throw a switch and the huge shutters begin to slide apart, revealing a darkening sky. I continue:

> Low down in the southwest Antares blinks a red-eyed sad farewell to fall while just above the horizon in the far northeast Capella sends flickering beacon flashes through the low bank of smoke and haze that hangs above the town. Instinctively I turn and look back toward the southeast for Capella's

"The world's greatest nonprofessional astronomer" is how Harvard astronomer Harlow Shapley described Leslie C. Peltier (1900–1980). A prolific observer, Peltier discovered 12 comets and two novae, and made more than 132,000 variable-star observations. His autobiography, *Starlight Nights*, celebrates the virtues of being an amateur stargazer.

co-riser. Yes, there it is, Fomalhaut, the Autumn Star, aloof from all the others, in a sky made darker by the rising purple shadow of the earth.

Leslie Copus Peltier, who was born January 2, 1900, and died May 10, 1980, in Delphos, Ohio, was already a famous stargazer decades before *Starlight Nights* first appeared in late 1965. Peltier saved $18 to buy his first telescope — a 2-inch brass refractor — by picking 900 quarts of strawberries on his family farm at 2 cents per quart. In 1918 he joined the American Association of Variable Star Observers, then a fledgling organization. On November 13, 1925, he discovered the first of his dozen comets, and six years later, when his brightest one glided gracefully across the sky, he was arguably the most famous amateur astronomer in the Western Hemisphere. By the time I began looking skyward in 1960, Peltier had become more involved with monitoring his beloved variable stars.

I first came across his name when I read in the March 1963 *Sky & Telescope* about his codiscovery of a nova in Hercules. A few weeks after the appearance of Ikeya-Seki, the Great Comet of 1965, *Starlight Nights* was published by Sky Publishing. I began reading it one evening and couldn't put it down. It was absolutely mesmerizing, this autobiography of a man whose lifelong passion was the night sky, its variable stars, and wandering comets. In public lectures I give these days, rarely does an event go by without my quoting something from this book. As we commemorate the centennial anniversary of Peltier's birth, Sky Publishing has reissued *Starlight Nights*.

For the first time, the new edition will have an index. While preparing it I had the most fun with the topics listed under "I" called "Insights." The book offers Peltier's wisdom on so many aspects of life that have little to do with astronomy and everything to do with living life to its fullest and appreciating the wonders of the environment around us. One such insight is about light pollution. When he noticed one night how all the fields around where he grew up were lit by bright lights, he wrote:

> The moon and the stars no longer come to the farm. The farmer has exchanged his birthright in them for the wattage of his all-night sun. His children will never know the blessed dark of night.

*Starlight Nights* is a book for every reader of this column, especially its young, city-bound budding stargazers who have never seen a farmer's field, a dark sky away from city lights, or the gibbous Moon rising over a distant mountaintop. There are so many titles out there that tell us *how* we should watch the sky and *what* we should do to get the most out of our telescope. *Starlight Nights* tells us *why*. This book won't teach you how to set up and initialize your computerized "Go To" telescope or process a CCD image, but it will make you stop and appreciate the beauty of what your telescope shows you. This book is about astronomy's big picture and it reminds us why we chose to become skywatchers in the first place.          FEBRUARY 2000

### Author Update:

Starlight Nights, the Adventures of a Star-Gazer, *is still available from Sky Publishing.*

# Telescopes for Telethon

When I was 14, I spent a year at the Jewish National Home for Asthmatic Children in Denver, Colorado. I remember the small buildings where we slept, the hospital where we went for treatment, and the night I was accused of using my telescope to look into the girls' dormitory window when I really was looking at Mars. But what struck me more than anything else during this difficult year was the inscription on the archway of the home's main entrance. "To help a child to health," it proclaimed, "is to walk with God."

In 1998 my wife, Wendee, and I decided to launch a campaign that focuses on a different group of childhood illnesses. Called Telescopes for Telethon, this annual event is designed to help raise funds to combat muscular dystrophy, a class of genetic disorders that results in slow but progressive degeneration of a child's nerves and muscles. While asthma is a serious, debilitating disease, its sufferers usually don't have the sadness of children with muscular dystrophy, which sentences them to shortened lives and every year brings a decreasing ability to move about, enjoy life, or even look up at the stars. Spearheading the effort to find its cure is the Muscular Dystrophy Association (MDA), a private, voluntary health agency based in Tucson, Arizona, known for its annual Labor Day Telethon hosted by entertainer Jerry Lewis. Established in 1950, the MDA aims to conquer the 40 neuromuscular diseases that affect more than a million Americans.

Wouldn't it send a message of hope to these children if astronomers worldwide could band together and show that we care by helping support the MDA's work? At least one day each year astronomy clubs, planetariums, and university astronomy departments can set up their telescopes and offer the public views of the night sky in exchange for a donation to the MDA.

In 1998 we hosted our telescope telethon on September 6th to coincide with Jerry Lewis's MDA Telethon. The following year we chose the first-quarter Moon weekend of June 18–19, since the Moon provides a wonderful observing target for any site, urban or rural. For this year's event we suggest either the night of the first-quarter Moon on Wednesday, May 10th, or Friday and Saturday, May 12th and 13th. We'd like you to participate, though, on *any* night near first-quarter Moon, when the Moon's phase resembles a "D" for dystrophy.

Last year Meade Instruments donated a 10-inch LX200 telescope to the group that raised the most money. The Fort Bend Astronomy Club near Houston, Texas, won the prize by raising more than $1,400. The club members set up their telescopes outside a theater complex and offered movie patrons views of the Sun. They conducted this session twice and collected even more funds during their regular monthly meeting! It

A youngster gets his first look at the Moon through a telescope during an astronomy summer camp hosted by the Muscular Dystrophy Association (MDA) last June near Tucson, Arizona. (That's me in the striped shirt.) My wife, Wendee, and I are currently helping the MDA to raise funds and awareness about the disease.

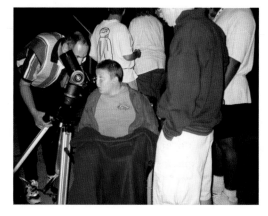

was so encouraging to see groups throughout the U.S. and Canada join the project with such vigor. This year Meade will donate another 10-inch LX200. It would be very nice if other telescope firms would follow Meade's lead and offer additional prizes to sweeten the pot.

In 1999 we learned that some clubs don't believe in charging anyone to look through a telescope. We understand that sentiment, but astronomical conscientious objectors can still help in an important way with Phase 2 of Telescopes for Telethon. The MDA operates a series of week-long kids' summer camps throughout the United States. We strongly urge you, through your local MDA office, to locate the nearest camp and organize with them a stargazing session with these children. Bring telescopes with eyepieces that can be accessed by youngsters from their wheelchairs.

On the night of June 23, 1999, Wendee and I drove to the mountains of northeastern Arizona, where we set up two small telescopes for a camp of 150 enthusiastic children and volunteer staff. We were overwhelmed by the energy of these kids as they got their first glimpse of the Moon, Venus, and Mars through the scopes. One boy spent some time expertly maneuvering his wheelchair to the best possible spot, then he looked at the Moon. "Is it all right," he asked, "if I just look for a while?" No one in the line complained as he stared and stared, trying to get the most out of his visual flight to another world. Did my telescope help that child to good health that night? I don't know. But for that glorious moment, Wendee and I, with all those children, really felt we were walking with God. MARCH 2000

### Author Update:

*For more information about muscular dystrophy, visit the Muscular Dystrophy Association's web site at www.mdausa.org.*

# A Marriage of Science and Art

Among the professional researchers I have featured in Star Trails to date, William K. Hartmann is unique. As a scientist, he is one of the originators of an unconventional yet widely accepted theory about the origin of the Moon. As one of the world's top astronomical artists, he has given us an idea of what other worlds are like. Finally, as a science-fiction writer, he has offered us a sense of what Martian life might be like in his 1997 novel, *Mars Underground.*

Born in 1939 in western Pennsylvania, Hartmann's early interest was awakened by two books — Chesley Bonestell's 1949 *Conquest of Space,* which was filled with Bonestell's magnificent paintings of space vistas, and an encyclopedia volume that contained a map of the Moon. "I was thrilled to find out that the Moon was actually a world with named craters, mountains, plains, and valleys," Hartmann recalls, "like a lost kingdom in the sky where no mortal had been." When he was 14 his father helped him build a 3-inch reflector. During high school in New Kensington he ground and polished an 8-inch f/9 mirror that became the heart of a telescope he still has to this day.

Anxious to begin a career in astronomy, Hartmann followed the standard route by studying physics. By the time he graduated from Pennsylvania State University in 1961 his mind was focused on planetary science. The challenge at the time was that, though the Moon landings were just a few years away, there was not a single planetary-science graduate program in the entire country! Fortunately, the Dutch astronomer Gerard Kuiper had just moved to Tucson to establish the University of Arizona's Lunar and Planetary Laboratory. So Hartmann decided to head west to pursue his graduate program there. In 1965 he obtained his master's degree in geology and, a year later, his doctorate in astronomy from the University of Arizona. He then went on to teach at the university for a year before joining what became the Planetary Science Institute, which now has offices in Arizona and California. In 1968 he met Gayle Harrison, an archaeologist and former Peace Corps volunteer. They married in 1970 and have a daughter, Amy.

The magic of the Moon map that so occupied Hartmann's teenage years never left him. In 1962, while he was a graduate student, he and Kuiper became the first to recognize the nature and structure of the multi-ring impact basins on the Moon, including the discovery of Mare Orientale's bull's-eye impact pattern along the celestial eastern limb of the Moon. How big, he wondered, could the largest impacts have been?

This question led Hartmann and his colleague at the Planetary Science Institute, Donald R. Davis, in 1975 to propose the "giant-impact hypothesis" to explain the Moon's genesis. They suggested that a giant

*Right:* Hartmann has rendered hundreds of space paintings, including this mural for the Chabot Space & Science Center in Oakland, California. *Below:* In 1975 planetary scientists Hartmann and Donald R. Davis proposed that a wayward body, perhaps the size of Mars, collided with the primordial Earth some 4.5 billion years ago, and that the debris from that cataclysmic impact quickly coalesced to form the Moon. This painting by Hartmann, based on later computer models, shows superheated terrestrial materials being ejected into space roughly five hours after the crash.

preplanetary body collided with the Earth some 4.5 billion years ago, blowing out rocky debris. A fraction of that debris went into orbit around the Earth and quickly aggregated to form the Moon.

"Part of the idea was to explain why the Moon has no iron core, even though the Earth has a large iron core," says Hartmann. "The Earth's iron had already settled to its center by the time the catastrophic impact occurred. Therefore the debris blasted out of both the Earth and the impactor came from their iron-depleted, rocky mantles." (The iron core of the impactor melted and merged with the Earth's iron core, according to later computer models.) This accounted for the lack of water and volatiles on the Moon (they were vaporized by the impact's intense heat), the overall similarity in composition of lunar and terrestrial rocks, the orientation and evolution of the Moon's orbit, and the Earth's relatively fast spin rate. The giant-impact hypothesis has since become the most widely accepted modern explanation for the origin of the Moon.

Hartmann's unique melding of art and science takes its roots from his childhood. "My grandfather, who came over from Switzerland, was a painter, and we had his landscapes around the house," he recalls. "So as a boy I always drew pictures as a form of entertainment." As a teenager, he built plaster models of the Moon's surface that he subsequently photographed. Thus, by 1955 he had his own imaginary "astronaut photographs" of the lunar terrain.

"I'm basically self-taught," says Hartmann. "I began to develop my painting more seriously in the 1970s when I realized I could use it to illustrate the astronomy textbooks that I was writing. I realized that it might make astronomy more interesting if students could visualize what it might be like to visit various astronomical locations." At the same time, he commissioned works from his artist friends and learned a lot from them. "In 1982 we organized the first International Space Art Workshop in Hawaii Volcanoes National Park," he says, "a gathering from which eventually grew the International Association of Astronomical Artists." Recently, Hartmann finished a 17-meter mural for the Chabot Space & Science Center in Oakland, California.

In 1985 Carl Sagan inscribed a copy of his novel *Contact* for Hartmann in which he described the latter as having a "functioning Corpus Calossum." Hartmann says with a smile that he rushed to a dictionary "to find out what Sagan meant or how he had insulted me." It turned out

Sagan was referring to the nerve bundle that connects the brain's two hemispheres. It was a compliment to a man so gifted in both scientific analysis and artistic passion. JUNE 2001

# The Last Hour

It's 3:30 a.m. The alarm clock starts ringing softly, and I quickly turn it off — I don't want to wake up my wife, Wendee. Quietly, I peek through the window blinds to make sure the sky is clear. The star-studded view that greets me confirms that it is. I walk out the back door and look up at a sky that's so dark I can dimly see the fuzzy speck that is Messier 33 in Triangulum.

It's the last hour of night before astronomical twilight begins, and for me it's an experience unlike anything else — the entire world seems perfectly at peace. The heat and haze that often mark the early-evening hours have dissipated, leaving the sky a lot darker, steadier, and crisper. All the neighbors are sound asleep, though tonight our beagle joins me for his nocturnal forage. Most nights, when I do my final hour of observing, it's just the sky, my telescope, and me.

On this particular early-autumn morning a tapestry of brilliant stars lines the eastern sky. It's the Heavenly G, a variation of the familiar asterism known as the Winter Hexagon. I can visualize the Heavenly G by tracing a line from Aldebaran through Capella, Castor and Pollux, Procyon, Sirius, and Rigel; the G is completed with a line going inward to Betelgeuse. Some observers see the pattern slightly differently: the G starts from Capella and goes all the way around through Aldebaran before ending at Betelgeuse. Spanning some three hours of right ascension and more than 60° of declination, the Heavenly G is the biggest and brightest asterism in the sky.

The peace and quiet that accompany my predawn adventure allow my mind to wander to nights of long ago. Sirius shines low in the southeast; five thousand years ago the ancient Egyptians timed the planting of their crops with its heliacal rising — the Dog Star's first appearance after its conjunction with the Sun — which usually heralded the annual flooding by the Nile River.

When the sky is this dark and serene, I feel as though I am standing next to, and learning from, observers of yesteryear who made their marks in the annals of astronomy. I imagine standing next to Clyde Tombaugh as he loads a 14 × 17-inch photographic plate into the 13-inch astrograph at Lowell Observatory near Flagstaff. Tombaugh chose Wasat, Delta (δ) Geminorum, as his guide star that evening in January 1930 when he

*Right:* Although it's not as well known as the Summer or Winter Triangle, the Heavenly G — a large star pattern consisting of the brightest stars of Taurus, Auriga, Gemini, Canis Minor, Canis Major, and Orion — is easily recognizable on crisp winter nights. *Below:* Dawn breaks and another observing session ends. I find the last hour of darkness before the start of astronomical twilight as the most peaceful time to reflect on and appreciate the beauty and majesty of the heavens.

made the first of three exposures. Three weeks later, he discovered Pluto lurking close by Wasat.

About 16° southwest of Wasat lies a strange, fan-shaped nebula, NGC 2261, which surrounds the variable star R Monocerotis. I see myself watching and learning from a young student named Edwin Hubble as he was preparing to take his first photographs of this 10th-magnitude nebula with Yerkes Observatory's 24-inch reflector in the winter of 1915. When he compared his pictures to images taken seven years earlier, he saw that NGC 2261's west side had expanded and become more radically convex. Who knows what would have happened if Hubble's interest had not been spurred on by what we now know as Hubble's Variable Nebula? Would he have returned to the Midwest to practice law? Or would he still have gone on to become one of the foremost observational astronomers of the 20th century and revolutionize our knowledge of the size, structure, and properties of the universe? That little nebula has a lot to answer for.

At the Texas Star Party I often share the predawn hours with some of the best deep-sky observers of our time. But I remember one particular incident when the sky had clouded over in the middle of the night and then cleared up again. I found myself surrounded by a field full of telescopes whose electronic beeping and whirring had fallen silent. Few people were still awake that morning. Instead of just me and a single scope it was me and dozens of scopes, most of them covered with tarps. I was concentrating so carefully in scanning the sky within the Great Square of Pegasus that I failed to hear the approaching footsteps. So I was surprised to hear a voice asking me what I was observing. I looked up, smiled, and explained, "I'm searching for comets." The guy looked at me as if I were nuts, shook his head, and laughed out loud as he walked away.

Not all interruptions are unpleasant, however. Just as I was finishing up on the morning of June 22, 1993, I saw a great cometlike flare in the east. I quickly turned my 6-inch reflector toward it. In the field of view was a brilliant "star" with a flaming tail trailing behind it. As I followed the star I quickly realized that I was witnessing a launch from the White Sands Missile Range in New Mexico, some 250 miles to the east. Less than a minute later it faded rapidly as the rocket's engine shut off. Suddenly there appeared in the field of view two groups of objects totaling 11 in all, racing across the sky in ballistic trajectories at a rate of a degree per second. Whatever these objects were, all of them were rotating — they were varying in brightness every few seconds before they finally disappeared behind the trees. Suddenly, a burst of greenish white light appeared, which spread out across the eastern sky and then faded. The experience reminded me of Tombaugh again, not at Lowell this time, but later in his life after World War II, when he developed optical-tracking telescopes for the U.S. Army to follow launches like this one from White Sands (see page 107).

Observing during the last dark hour of the night is not an escape from the realities of life; it's a celebration of it. And I think we're losing something precious every time we pass up the chance to participate in the human interaction with the sky that comes from being outside, alone with our telescopes on a clear, starlit night. JANUARY 2003

# The New Age of CCD Observing

In the last decade the rise of commercial CCD (charge-coupled device) cameras has been nothing short of astonishing. Although still relatively few amateur astronomers are using CCDs right now compared to film, the dramatic increase in the number of camera accessories, books, software, observing projects, meetings, and Web sites devoted to CCD

In our digital age more and more young people are getting introduced to the technological wonders of CCDs, taking and processing electronic images with their parents and at schools, star parties, and astronomy camps.

imaging clearly shows that interest in this new observing field is growing exponentially each year.

Although I was among the first amateurs to enter this brave new world, I still approach it kicking and screaming. Back in September 1985, when I joined the International Halley Watch, I worked with planetary scientist Stephen M. Larson at the University of Arizona's Lunar and Planetary Laboratory near Tucson. There we used a huge, bulky CCD camera system at the focus of the university's 61-inch telescope to study Halley's Comet, which was making its first comeback since 1910. Our camera's detector is crude by today's standards, offering an array of only 320 by 512 picture elements, or pixels. (I knew so little about how CCDs worked at the time that I remember running to a friend's office nearby to ask with some embarrassment, "What's a pixel?")

To achieve greatest sensitivity the detector had to be cooled with liquid nitrogen ($LN_2$). So before each observing run Larson and I stopped by a cryogenic supply facility to fill our Dewar flask with this supercold –196°C (–321°F) liquefied gas. Preparations for observing back then weren't easy — the computer stack needed to operate the CCD camera was very heavy, so it had to be carted about on a large heavy-duty dolly. There was another dolly to carry the $LN_2$ tank and a third for the camera itself.

I was curious to see how our CCD system would work with Minerva, my homemade 6-inch f/4 Newtonian reflector, and so on December 17, 1988, I tried it at home. Since the camera's weight was at least 10 times that of my little scope, I couldn't mount the camera directly to the scope's eyepiece holder. Instead, I braced the camera using several encyclopedia volumes and aimed it at Minerva's eyepiece. I then propped up the scope's short tube using more books and pointed the whole setup toward Polaris. I took a couple 45- and 90-second exposures. When I displayed the images, I found that my short exposures were able to record stars fainter than 14th magnitude. They also captured tiny galaxies that I couldn't even see in the same field visually. Amazing!

Today's CCD systems are light, compact, and portable — they can be inserted right into the eyepiece holder, and their output cable can be plugged into a laptop computer. The detectors are now also cooled thermoelectrically instead of using $LN_2$. Not only that, but the advent of computerized Go To telescopes and automation software now allows you to remotely command your telescope setup to find an object to an accuracy of less than an arcminute, take a CCD exposure, calibrate the image, save it, and move on to the next target. You can repeat this process over and over throughout the night. If you have a dome observatory, there are programs to control the dome's rotation to match the movement of the telescope across the sky. And for those who want to find new asteroids, comets, or supernovae, there are programs that can show in your images any object that moves, brightens, or fades.

It's clearly technology that has allowed this progress to happen, technology that's now trickling as far down the educational ladder as elementary schools. Can you imagine a 6th-grade boy or girl having the opportunity to discover an asteroid on his or her first night? If projects led by young people such as Ryan M. Hannahoe are any indication, this will happen soon. Hannahoe, who chairs the Astronomical League's Youth Activities Committee, has teamed up with Denver University astronomer Robert Stencel to form the Virtual Telescope Program. The program currently gives high-school sophomores and juniors a chance to experience hands-on, interactive, remote CCD imaging via the Internet using the large telescopes of participating professional observatories.

There's no doubt that this is the way amateur observing is heading. But should it be the only way? By the time a student reaches college, he or she should certainly have mastered this new technology. But elementary school? Is there no more room for a romantic look at the night sky through a small telescope, like the view I got through my 3½-inch reflector back in 1960? I'll never forget my first glimpse of Jupiter that warm summer evening. It was unique; not an image of the planet on film or on a computer screen like a TV show, but the real thing — photons traveling nearly an hour through interplanetary space, bouncing off the telescope mirror's aluminum coating, and then tickling my retina.

No matter how advanced the CCD revolution becomes, I think there will always be room for amateur skygazers to enjoy a starlit evening, finding their favorite celestial gems by traditional star-hopping or recording their views with old-fashioned drawings. The human eye is a unique optical wonder. It may not have the quantum efficiency of a CCD detector, but it does have the ability to discern subtle details and a wide dynamic range of brightness and contrast that no camera can. Its main drawback is that it's not as easy to quantify accurately what the eye sees as it is to measure what a CCD records. A skilled visual observer can estimate the magnitude of a variable star consistently to a tenth of a magnitude; an amateur with a CCD typically can measure it to a hundredth.

There is something, however, to be said about having to crawl first before learning how to walk. CCD observers can, if they like, do it all by the push of a button or the click of a mouse, never having to learn the whereabouts of a single constellation. I'm not in favor of one form of observing over the other — I enjoy both — but I hope that, especially for young people, the era of visual observing, with all its magical vistas, challenges, and limitations, never ends. MARCH 2003

# In the Footsteps of Giants

My dad loved to tell the story of how the outermost worlds of our solar system — Uranus, Neptune, and Pluto — were discovered. His account inspired my young mind: in 150 years the size of our solar system had increased four times. We all know how it began, with William Herschel's accidental discovery of Uranus in 1781, which, in turn, led to the mathematical prediction of the position of Neptune by John Couch Adams and Urbain-Jean-Joseph Le Verrier in the 1840s, and finally to Clyde Tombaugh's successful search for Pluto in 1930.

These raw facts don't do justice to these people, who struggled and fretted over their discoveries. Of the four, Herschel had it the easiest. The story of Neptune was far more complex — by the 1830s astronomers were starting to worry about Uranus's persistent wandering from its predicted path, which couldn't be explained by the gravitational influence of the Sun and the known planets. In the early 1840s Adams, then a young student at the University of Cambridge, proposed that a massive, more remote planet was perturbing Uranus. In 1845, two years after he graduated, he sent his calculations of where this new planet might be to Cambridge astronomy professor James Challis, who then handed his work to England's Astronomer Royal, George Biddell Airy, at Royal Greenwich Observatory. Although Airy seemed interested, he did nothing about ordering a search.

At around the same time Le Verrier tackled the Uranus problem independently and published his own calculations of the new planet's location. A copy of his paper reached Airy, who commented that the young French mathematician's prediction agreed to within 1° of what Adams had predicted. Airy finally instructed Challis to search for the object. But two things went wrong: Airy neglected to tell Adams about Le Verrier's work, and Challis, instead of going right to the predicted position, mounted a cumbersome star-by-star visual search over a large area of the sky, going over the new planet twice without recognizing it.

Nor did Le Verrier have any luck in getting a search going at Paris Observatory in his native France, so he went to Johann Galle at Germany's Berlin Observatory. Intrigued and excited, Galle, assisted by Heinrich d'Arrest, began a search at Le Verrier's predicted position and found the new planet the very next night!

After the discovery was announced, Airy tried to have Adams credited with it as well. The French were livid with this belated attempt to get some recognition for England. Although an international dispute broke out, most sources today give Adams and Le Verrier equal credit for the role they had played in finding Neptune.

The planet saga continued with Percival Lowell mounting his search for Planet X at Lowell Observatory in Flagstaff, Arizona, finally culminating in Tombaugh's discovery of Pluto. In 1963 I met Clyde at a convention in Denver. It was a highlight of my life — meeting a real-life planet discoverer. Even after writing Clyde's biography,

Herschel wasn't expecting to discover a new planet, and he first thought the world he'd found was actually a comet. Herschel was, however, the first person since antiquity — and the first person whose name we know — to recognize a new planet. For that reason he is honored as one of history's six greatest astronomers on the Astronomers Monument in front of Griffith Observatory in Los Angeles.

I never thought I'd get close to following in the footsteps of Herschel, Adams, and Le Verrier. But in December 2002 that began to change.

While on *Sky & Telescope's* eclipse tour to Africa, our group visited the Cape Town site of William Herschel's great 20-foot-focal-length reflector, brought there by his son, John, in the 1830s for the first survey of the southern sky. An obelisk marks the site, now situated on the grounds of an elementary school. A few months later I encountered Herschel's actual telescope; its tube and optics were on display at Smithsonian's National Air and Space Museum in Washington, DC. The scope is set up on a model of Herschel's mount, along with figures of William peering through the eyepiece while below him, his sister, Caroline, takes his observing notes.

And then last September, as I was on my way to give a lecture in Bristol, England, I couldn't resist stopping by 19 New King Street in nearby Bath, the home of William and Caroline. This time I felt as though I were visiting old friends. Just as I have enjoyed hearing Clyde and his wife, Patsy, tell me of their adventures regarding Pluto, here I was ushered into the Herschel house to hear, vicariously at least, William talk about his planet. I imagined him sharing with me the congratulatory letter he received from comet hunter Charles Messier, and then Caroline musing how searching for comets might be a worthwhile pastime. Turning toward his sister, William would say with a smile: "Lina, yes! I think you could find a comet someday." (Caroline did find the first of her eight comets in 1786, after they had moved to a place near Windsor Castle.)

It's extraordinary how a visit to some key places can actually be a substitute for "meeting" a person you would never get a chance to see. But for me, the strongest connection came not inside the Herschel house but outdoors in the garden, where William had set up his 7-foot (focal length) reflector to conduct his sky survey and, on March 13, 1781, discovered the solar system's seventh planet. As I looked up in the northwest, I mentally placed the star H Geminorum in the sky and imagined hearing Herschel's words:

> While I was examining the small stars in the neighborhood of H Geminorum, I perceived one that appeared visibly larger than the rest; being struck with its uncommon magnitude I compared it to H Geminorum and the small star in the quartile between Auriga and Gemini, and finding it so much larger than either of them, suspected it to be a comet.

I left the Herschel museum struck by this experience. I knew Clyde personally, and now I felt I knew the Herschels. But aside from seeing Le Verrier's statue on the lawn of Paris Observatory, I wondered if I would ever have a similar experience with the other discoverer of Neptune. As it turned out, I didn't have to wait long.

A few days after my visit to Bath, I was at the University of Cambridge library perusing *Euclides Elements of Geometry*, a 17th-century book whose preface was written by John Dee, Queen Elizabeth I's famous science adviser. Among its features was a gem of an introduction in which Dee makes the startling observation that "the distance of the starrie skie is from us, in Semidiameters of the Earth 20081½. Twenty thousand four-score, one, and almost a half." How he arrived at this odd figure is beyond me, but there is one thing I am sure of: the English discoverer of Neptune must have wondered the same thing. As I closed the book, its bookplate caught my eye and sent me into orbit — Ex Libris John Couch Adams. I was sharing a bit of insight about Dee with a codiscoverer of Neptune!

The following morning I visited the Institute for Astronomy at Cambridge, where Adams served as director for many years after his discovery. I stood for a long time before the beautiful 11.6-inch (29-centimeter) Northumberland refractor with which Challis had tried in vain to search for Adams's planet.

Clyde and I often reminisced about the Neptune controversy. "I don't know why Adams just didn't set up a telescope," he told me once, "and look for the planet himself!" After all, Neptune was then shining brightly at magnitude 7.8. Some questions might just have to remain as such; after all, there are no what-ifs in history.

The discovery of the three outermost planets is one of the most adventure-laden narratives in the history of astronomy. What adds color to the story, though, are the interesting personalities of its main players. Herschel, Adams, Le Verrier, and Tombaugh all were decent people with an undying enthusiasm for celestial exploration. They also shared something else, said Clyde: "the imagination to recognize a discovery when they made one."                                         FEBRUARY 2004

# Don't Let It Get to You

A few months ago I encountered a teenager whose overzealousness had caused him some trouble in an astronomy club. Worried that the problem might cause this enthusiast to withdraw from astronomy, I recounted to him how a similar thing happened to me — *twice* — when I was young. When I was 13 and again six years later, two events almost made me give up my love and passion for the night sky.

In 1961 the Royal Astronomical Society of Canada's Montreal Centre was led by the late Isabel K. Williamson. An inspiring leader with a lot of enthusiasm, Miss Williamson did have a harsh side. At that time the Centre was planning "Star Night," a stargazing event involving some 20 telescopes for an expected crowd of more than 2,000. So, late that Saturday evening, I offered to assist. "No," Miss Williamson told me.

No? Just no? After all, my telescope and I had been through quite a bit in those past few months. One frosty April evening, when I placed my scope's cast-iron tripod into a packed snowdrift, the icy platform cracked, sending the telescope and me into a snowbank. I had wanted to display my freshly repaired instrument. But this time I suggested to Miss Williamson that if I couldn't bring my own scope, perhaps I could assist one of the other members with his or her instrument.

"Listen," Miss Williamson glared at me. "We have good reasons for not allowing anyone to join our organization until they're 16. You will not bring your telescope to Star Night, and you will not assist anyone else."

Even now I remember the unbelievable feeling of desperation that came over me as I listened to her words, coming from a mentor I highly regarded. On the bus ride going home that night I decided to drop out of astronomy. At the time I told no one — neither my friends nor family — of the incident, even though it might have contributed to the massive asthma attack a month later that hospitalized me for two weeks.

By the time I was out of the hospital I was back in astronomy again, writing my teenage book, *An Encyclopaedia of the Universe*, in every off-school moment. (Vastly revised, that book recently was published as *Cosmology 101*.) I began my comet hunt in 1965, and Miss Williamson appointed me co-chairman of the Centre's Comet and Nova Search section the following year. Astronomy provided great satisfaction until April 15, 1967, when I discovered that the Centre's wind-up barograph had stopped recording air pressure on its sheet of paper. The next night I mentioned this to Miss Williamson, an act that almost cost me my RASC membership.

At the end of that meeting she asked me why I had overwound the barograph. My denials made matters worse, and within 30 minutes I was ordered to leave the building. A month later, when I tried to return, I was told to leave once again, and this time I was pushed out the door by three people as the rest of the group stood by and watched. The following day, in an emergency meeting of the Centre's Board of Directors, a motion was put forward to have me expelled from the Royal Astronomical Society of Canada. If it hadn't been for friends like Constantine Papacosmas, I would probably not be a member today.

Some 11 years later I decided to write to Miss Williamson. She replied that she was well, and as our new correspondence continued, she invited me to visit. This started a tradition of visits each time I returned to Montreal until her death four years ago. She congratulated me on my award of the RASC's Chant Medal, a medal she herself had won in 1948, and as we chatted she revealed that she had never won the organization's National Service Award. I quietly nominated her for it, and it was presented in 1981. A few years later, I proposed that the Centre's meeting place be named the Isabel K. Williamson Observatory.

I'm now the Montreal Centre's honorary president. Years ago there was no youth section like what the Astronomical League now has, nor did the Astronomical Society of the Pacific have its Project ASTRO, which goes into classrooms around the US. But these programs don't address the interpersonal contacts and unplanned events that can help set the course of young people's lives. As I had suggested to that overzealous teenage astronomer, don't ever let anything, particularly a slap on the wrist from a club officer or board member, keep you from a lifetime passion for the stars. I'm glad to say that he did heed my advice. JULY 2004

# Earth Strikes Back

It happened about 25 minutes after Ernst Wilhelm Leberecht Tempel's comet was struck by Deep Impact's 372-kilogram (820-pound) copper projectile at 10:52 p.m. local time on July 3rd. Our group — my wife, Wendee, and I; Thom and Twila Peck; John and Liz Kalas; and Michael Terenzoni — had been observing Tempel 1 from my backyard observatory in Vail, Arizona. Our eyes were glued to the eyepieces, watching for any visible changes in the comet's appearance. At impact time, NASA TV showed a brilliant flash of light emanating from Tempel 1's nucleus, but that particular cosmic fireworks display lay buried deep within the comet's dense, fuzzy inner coma, and we didn't see a thing.

Time passed. After about 9 minutes I thought I could detect some brightening in the coma but later admitted that I could just be dreaming. After another 5 minutes, my friend Bob Summerfield called. "Remember the infrared image of Comet Shoemaker-Levy 9 hitting Jupiter, with the whole planet lit up?" he asked. "On TV the entire bottom half of the comet's nucleus was lit up just like that. It was as if a huge flashbulb went off!"

Still we waited. And then, about a half hour after impact, the small, distant light from the comet soon became obvious. From her vantage point at our 16-inch reflector, Wendee exclaimed, "I think I see a pinpoint of light in the comet." It wasn't much — a 12th-magnitude "star" near the center of an 11th-magnitude fuzzy blob. But it effectively doubled the comet's central brightness in an instant. It was hard to watch the unfolding drama in a world almost as far from us as the Sun without appreciating the history of the moment: this comet was reacting to something we humans had done to it. After a planetary lifetime of cometary bombardments, Earth struck back. As Deep Impact lead scientist Mike A'Hearn wrote to me, "It's revenge of the dinosaurs!"

During the past two decades, three robotic space probes have made close flybys of comet nuclei: Giotto flew by the heart of 1P/Halley in 1986, Deep Space 1 spied 19P/Borrelly in 2001, and Stardust passed by 81P/Wild 2 in 2004. Deep Impact's visit, however, was a first. It was not just a visit; it was a definite statement from Earth.

The idea for the mission dates to 1996, when engineer Alan Delamere and astronomer Michael J. S. Belton proposed it to NASA. "Comets are exceptionally diverse," Belton told me, "even more than previously believed. As the first contact mission, Deep Impact is a rather blunt tool — but now the way is open to new and more refined approaches to this kind of experiment."

Throughout human history, the appearance of brilliant naked-eye comets has always evoked fear, wonder, and admiration. In 44 BC, not long after Julius Caesar's assassination, a bright comet shone over Rome. Centuries later, it inspired Shakespeare in his tragedy *Julius Caesar* to ascribe these words to Calpurnia, Caesar's third and last wife, who had a premonition of her husband's death: "When beggars die, there are no comets seen; the heavens themselves blaze forth the death of princes."

"I have an ancient Roman silver coin with the portrait of Augustus Caesar on one side and the comet of 44 BC on the reverse," says mission scientist Donald K. Yeomans of NASA's Jet Propulsion Laboratory. "Deep Impact was designed to probe a comet's subsurface materials, to remove some of the mystery surrounding these objects, and, perhaps, to dispel — once and for all — the superstition that has surrounded them. Yet, for luck, I did have my Roman coin in my pocket the night of the spacecraft's successful impact on Tempel 1."  NOVEMBER 2005

*Left:* I obtained this image of Comet Tempel 1 moving from upper right to lower left, brightening obviously at impact. *Above:* Deep Impact's flyby spacecraft captured this view shortly after its impactor slammed into Tempel 1's 7.6-kilometer-long nucleus.

### Author update:

*The Deep Impact mission produced many surprises. Water ice was found in three small areas on the comet's surface, the first time ice has been detected on the solid core of a comet. Debris flung out by the collision contained carbonates and clays, two classes of minerals that on Earth form only in the presence of water. Numerous organic molecules were also discovered — molecules that are abundant in interstellar space but have not been positively identified (until now) in comets.*

# Four Decades of Comet Hunting

Forty years have passed since December 17, 1965, the night I began searching for comets from my backyard in Montreal. Three factors contributed to my teenage decision to undertake this search, which would eventually change the course of my life. The first was that I was well along in my quest to find and observe all the Messier objects (I had already found 75 by October of that year), and I was starting to plan what to do next after I completed the list. The Royal Astronomical Society of Canada's Montreal Centre offered a Herschel program, but I was looking for something a little different. Second was the discovery of the great sungrazing comet of 1965 by Japanese comet hunters Kaoru Ikeya and Tsutomu Seki (see pages 76 and 80). Their inspiring stories thrilled me, and while I harbored no illusions about the challenges of actually finding a new comet, I saw the search itself as a way to learn the night sky in a fun and unique way.

The third factor confirmed my decision after I had made it. Leslie Peltier's autobiography, *Starlight Nights*, the moving yarn of an Ohio amateur astronomer who discovered 12 comets, remains an exhilarating book and a source of inspiration (see page 20). Thus, my comet-search program was one of exploration, discovery, and education. As of December 2005, the program has achieved the following results.

**Becoming intimately familiar with the sky.** Over the course of my comet hunt I've encountered thousands of deep-sky objects, ranging from unusually red variable stars to distant clusters of galaxies. I picked up my first one, the open cluster/emission nebula NGC 1931 in Auriga, on January 1, 1966, and decided then to keep a list of the more interesting objects I came across during my search. That Levy List is the basis for my latest book, *Deep Sky Objects: The Best and Brightest from Four Decades of Comet Chasing* (Prometheus Books, 2005). The hundreds of objects sampled in the list include Levy 384, a possible new 11th-magnitude open cluster in Puppis. In my slow but deliberate way, with each evening or morning's comet search, I've tried to absorb what the sky wanted to teach me.

**Discovering and confirming Comets, Asteroids, and Novae.** To date, I've found 8 comets visually from my backyard in Arizona and 13 photographically while working with Gene and Carolyn Shoemaker at Palomar Observatory. Each comet is a highlight, of course, but of all these discoveries, four stand out. Comet Levy-Rudenko (C/1984 V1) was my very first find, and I spotted it on November 13, 1984, exactly 59 years after Peltier's first comet and on the same date (notwithstanding the change from Julian to Gregorian calendars) as Tycho's first observation of the comet of 1577. My second memorable comet was C/1988 F1, an intriguing cosmic interloper that is believed to have split from a larger body some 12,000 years ago. The Shoemakers discovered the other half a month after I spotted C/1988 F1. This was the first case, other than sungrazers, of pieces of the same comet being found separately. My third memorable one, C/1990 K1, put on a nice performance in August and September 1990, brightening to 3rd magnitude as it cruised the Milky Way and growing a broad antitail

the following year.

The best known of my comets, Shoemaker-Levy 9 (D/1993 F2), collided with Jupiter in the summer of 1994. It provided a wonderful opportunity to educate the world about the importance of

Here I am with my 4-year-old grandson, Matthew Vigil. Shown at right is the 0.3-meter (12-inch) f/2.2 Meade LX200SC Schmidt telescope that I use to look for comets with a highly modified SBIG STL-11000M CCD camera. To date, I've spent a total of about 3,100 hours searching for comets.

seeking out comets and asteroids that could someday threaten Earth.

Finally, I've found 282 asteroids so far, 156 of them while observing at Palomar and the rest with Canadian amateur Tom Glinos. I've also had two nova confirmations. On September 12, 1978, not far from the Adirondack Science Camp in Lewis, New York, where my wife, Wendee, and I now hold an annual astronomy retreat, I made an independent discovery of Nova Cygni 1978. Thirteen years later, I verified a "nova" that Clyde Tombaugh had discovered in 1931 in Corvus. It turned out to be a new cataclysmic variable star, now designated as TV Corvi.

**Embarking on a Research Program.** At 17 years of age in 1965, I hadn't a clue what a "research program" meant, but it seemed so lofty that I wanted to do it. It began with simply reading about comets and taking notes about what I read. These notes are still in my binder, along with those I later took for my master's-degree thesis about Victorian poet Gerard Manley Hopkins and his observations of the comet of 1864. Today, my doctoral thesis on the comets, novae, and eclipses of Shakespeare's time is an extension of that high-school passion.

To look at the past is helpful if one uses it to improve his or her future. After the Shoemakers and I ended our program at Palomar in 1996, Carolyn, Wendee, and I continued the search using film and a pair of 8-inch Celestron Schmidt telescopes. Our project gradually evolved into a CCD comet search, this time using a highly modified SBIG CCD camera coupled to a 12-inch Meade Schmidt telescope. It's been nearly 12 years since I've discovered a comet, but I'm still looking.

So, like Sir Joseph in the opera *HMS Pinafore*, I'd advise people who aspire to become comet hunters: "Now landsmen all, whoever you may be . . . Stick close to your desks and never go to sea!" In this day and age of automated professional sky surveys such as LINEAR, your chances of actually discovering a new comet in the cosmic sea are quite slim. But if you do catch the passion, I know of no better way to become acquainted with the treasures and the majesty of the night sky.     MARCH 2006

### Author update:

*On the morning of October 2, 2006, I discovered comet C/2006 T1 (Levy) from my Jarnac Observatory in Arizona. This was my first visual discovery in 12 years, and it brings my total of comet finds to 22.*

people

# Uncaged Spirit

From the outside, the only thing pleasing about the Federal Correctional institution — a medium-security prison in Tucson, Arizona — is its maze of low-pressure, sodium lamps. But inside, where some 750 inmates serve time for drug-related offenses, a few dozen have united to form our nation's only prison astronomy club.

Frank Lopez, owner of the Astronomy Shop in Tucson is familiar with the facility and its inmates, having been there six times in the last year. He goes there to discuss the night sky with the prison's astronomy enthusiasts. Lopez offered his services after receiving a call from Sean Williams, a recreational specialist at the facility; Lopez's business was the only astronomy number Williams noticed in a local telephone book.

On his first visit, Lopez entered the gates with trepidation. But the 25 club members he met soon eased his tensions. "Just think of us as regular guys who have made mistakes," one told him. Lopez found them serious about learning. With some of its recreational funds, the prison purchased a small telescope for the club and later replaced it with a computerized Meade LX-200 telescope.

On February 28th Dean Ketelsen, president of the Tucson Amateur Astronomy Association, and I joined Lopez for a night's program at the prison. I met Ketelsen at the prison's front gate — a formidably tall fence studded with coils of razor wire. Before entering, we had to remove our watches, belts, and shoes, and parade through security. Then someone stamped our hands with ultraviolet code so we could get out later.

Tim Gillooly, the new recreational specialist supervising the astronomy group, escorted us through the facility and expressed how proud he was of the administration's attitude toward the club. The grounds were spotless, and clusters of inmates in green jackets walked freely but aimlessly about; others had escorts. We arrived at a small but modern lecture room with state-of-the-art, audio-visual equipment. Gradually the room filled with some 20 inmates.

Our program lasted about two hours. Ketelsen discussed his work with the newly spin-cast, 6.5-meter mirror destined to upgrade the multiple-mirror telescope atop Mount Hopkins in Arizona. And I talked about my work on comets and introduced Minerva, my 6-inch comet seeker. Some of the inmates' questions were as incisive as those we might ask at our own club meetings: Does the 6-meter oven spin before the glass in it has completely melted? How do you report and name a comet?

We in turn learned how proud the inmates are of their club, the brainchild of one member who became interested in the stars while serving his 12-year sentence. An inventive person, he put his all into making it work. Why? "Because the stars are my ticket out of here," he said philosophically. "When I sit out in the yard and look up, I feel free again."

There were also lighter moments during our conversations. For example, when asked if anyone had anything to add, one inmate smiled and said, "Yes. I was framed!" And to a question about their plans for expanding the group, another suggested "field trips!"

With a clear sky two nights later, Ketelsen, Lopez, and I returned to the prison for an observing session in the compound's southeast corner. Aside from meeting weekly, the club members try to observe about eight times a month on Tuesdays and Thursdays. For our gathering, Ketelsen set up his World War II Japanese battleship binoculars, the club had its Meade LX-200, and I readied Minerva.

When asked to demonstrate comet hunting, I sat beside Minerva and began methodically sweeping the western sky, picking up several rich star fields, a Messier object, and a bright sodium lamp. During this exercise, one inmate turned his gaze skyward as if contemplating. He then mentioned the imminent state execution of convicted killer John George Brewer, which I had forgotten was to occur that night about 70 miles to the north in Florence. The group's liveliness temporarily disappeared. The silence broke when an inmate startled me by commandeering Minerva to search the skies for comets.

Another member tried finding Periodic Comet Schaumasse, but was defeated by light from the nearby Moon. The biggest hindrance, though, was the overabundance of artificial light. To reduce its impact, the club's founder designed a cyclopean hood, with a hole for the observing eye. He also assembled a long cap for the telescope to further reduce light-pollution effects. In the end, we got good views of the Moon, Venus, and the Orion nebula — objects found not by us but by the club members themselves.

Teaching astronomy in Arizona's prisons is a tradition of sorts. It began in 1966 after Bart Bok arrived in Tucson to direct Steward Observatory. Of the many letters waiting on his desk, one from the state prison in Florence piqued his curiosity. Soon Bok and his wife, Priscilla were deeply involved in the prison's Women's Self-Improvement Society. After dinner with the inmates, Bart talked to them about the Milky Way. "The question period was very structured," he recalled. "One of the prisoners chaired the meeting, and she would field the questions in order. And woe to anyone who asked a question out of turn. The chairperson explained, 'We are preparing ourselves for reentry into society.'"

The prisoners enjoyed this evening so much they invited the Boks back the following year. This time Priscilla sat next to someone in her mid-60s, "a very elegant woman," she later told Bart, "who had traveled widely. I can't imagine why someone with her upbringing, her class, would even be in a prison."

Priscilla's new friend turned out to be the "trunk slayer," Winnie Judd, among the most notorious murderers in Arizona's history. On October 16, 1931, Judd fatally shot two women, stuffed their bodies into trunks, and shipped them off to California. A baggage handler discovered the crime when he saw blood dripping from one of the trunks.

In other years Priscilla met with Judd and had animated conversations. Priscilla was so strongly attracted to this woman that she insisted that Bart write then Arizona Governor Jack Williams to intervene on Judd's behalf and commute her sentence. "If we ever need a housekeeper," Priscilla said believingly, "I would rather have Winnie Judd than anybody else. "Judd was released in the fall of 1970, a few months after Bart's letter. The Boks were not graced with her services, however, since she moved out of state.

Why share the night sky with prison inmates? Spreading the word about astronomy is a good idea for any audience, The stars are there for *all* to see. A prison might cage the body but not the soul.     JULY 1993

### Author Update:

*This was actually my second visit to a prison. The first was Millhaven Maximum Security Penitentiary in Ontario, Canada, where I talked to the inmates about the 1979 total eclipse of the Sun.*

# A Toast to Friends, Present and Absent

When the Moon's shadow rushes across a hairline swath of Earth, the eerie atmosphere it creates brings out deep emotions. Some people like screaming; others become quiet as they watch the fabulous interplay of Sun, Moon, Earth and themselves.

On May 10, 1994, at 8:42 a.m. Mountain time, I watched the dark Moon take its first tentative bite out of the Sun. The long-awaited annular eclipse had begun. I was close to the centerline and at a very special site: the New Mexico home of Clyde and Patsy Tombaugh. "We have a ringside seat!" Clyde punned.

Watching the annular eclipse with us was a diverse group of friends who mean an awful lot to me. First was Clyde himself. In recent years he has become a hero to many amateur astronomers, and with good reason. His story is a fairytale success. Seven decades ago, using the 9-inch reflecting telescope sitting with us at the eclipse site, he had made planetary drawings so fine that in 1928 the director of Lowell Observatory hired him to conduct a trans-Neptunian planet search, and in 1930 he discovered Pluto. Now 88, Clyde was enjoying the start of this eclipse. He had spent the past few days finishing a special eclipse telescope — a piece of wood, two unsilvered mirrors, and an eyepiece.

His old friend Brad Smith, who headed the imaging team for the Voyager project, was with us. Years ago, while a young scientist with the U.S Army Map Service, Brad had been assigned to take a clandestine look at one of Clyde's programs. Brad was so impressed that he eventually left the army to work with Clyde, who is proud of his coup. "He spied on me," Clyde notes, "and I kidnapped him!"

Jim Scotti and his family were also with us at the Tombaugh's home. Jim was the first person I met after moving to Tuscan in 1979, and I've watched his career move forward in leaps and bounds since he began working with the Spacewatch project in 1982. Its purpose is to find asteroids in orbits that cross the orbit of the Earth. "Find them before they find us!" is the Spacewatch battle cry. To the basic Spacewatch program Scotti added recoveries of comets, tallying his 45th just before the eclipse.

Just after he confirmed the discovery of Comet Shoemaker-Levy 9 in the spring of 1993, Jim told our friend Wieslaw Wisniewski about it. The Polish astronomer immediately took a stunning CCD image of the comet. Wieslaw was an indefatigable observer who thought nothing of running off to a remote observatory on a 30-night observing session. He died suddenly a few months ago. We miss him a lot, and as the Moon covered more of the Sun that day, we wished Wieslaw had been there to share it with us.

The annular eclipse of May 10, 1994, was visible along a path stretching from Texas to Ohio and eastward through New England. The images in this sequence were taken at five-minute intervals.

The eclipse had reached a critical stage. The Sun's light has become so reduced that the landscape looked eerie. Springtime eclipses have the added bonus of birds in fine voice, and when their singing dies down the effect is really pronounced. With a thousand crescent Suns projected on the ground through spaces between leaves, it appeared as though nature itself had become a poet.

Brad and Clyde started talking about other eclipses they had shared. One was a lunar eclipse seen from Lowell Observatory; they used the darkness to search photographically for natural satellites of the Moon. But they didn't find any "grandchildren of the Sun," as Clyde called them.

Soon the tiny Moon closed in on the Sun's opposite edge. For a few seconds the edge broke into an arc of Bailey's Beads, as the tops of lunar mountains sawed their way through the last bit of solar photosphere. Then came the spectacular ring. Although I had seen annular eclipses before, I had never been right at the center of one. A few minutes after the onset of annularity, the ring of sunlight was absolutely perfect.

The scene was surreal. It was a clear spring morning lit by a ring, in the sky so dark that Venus was easy to spot. Beneath the trees the shadowed ground was punctuated by hundreds of perfect rings. Suddenly, with another display of beads, the circle was broken.

As the sky started to brighten, Brad and Clyde recalled another eclipse they had watched together. It happened 31 years ago, on July 20, 1963, in Alaska. Braving mosquitoes and a dismal weather forecast, they set up their equipment and viewed a beautiful total eclipse. While listening to them, I remembered watching the same eclipse with my parents, a whole continent away.

I remembered how much fun that total eclipse had been, and how interested Dad had been in the orbital mechanics that led to such a poetically pleasing event. Even though he has been gone for 10 years, I suddenly missed him intensely. He was not there to enjoy either the eclipse in May or the comet crash that followed it in July. Eclipses do invite raw emotions, and even though I was among such first-class friends, I really missed the absent ones.  OCTOBER 1994

# Something Old, Something New

Ever since I got a 99 percent score on my seventh-grade spelling test on March 23, 1961, that date has had special meaning in my life. On March 23, 1931, Clyde Tombaugh found a new cataclysmic variable star in Corvus; on March 23, 1990, I made the first visual observation of that same star — now named TV Corvi — in outburst. On March 23, 1993, Eugene and Carolyn Shoemaker and I took the two discovery photographs of Comet Shoemaker-Levy 9. Now we're in 1997, a year that had a deep partial lunar eclipse on March 23rd.

For about six years, my mother had been urging me to write to her friend's daughter. On March 23, 1992, I finally wrote to Wendee Esther Wallach, a physical education teacher in Las Cruces, New Mexico. At the time I was visiting Clyde at his place in Las Cruces every month to interview him for his biography. By the time Wendee and I finally met, I was working on the book's third printing.

On May 10, 1994, we watched the annular eclipse of the Sun from Las Cruces. While I was with a group of friends at Clyde's house, Wendee had her P.E. classes view the event using a variety of safe methods. Everybody had a great time — it was one of the most enjoyable eclipses I've ever experienced. In April the following year, when the citizens of Las Cruces honored Clyde and his wife, Patsy, at a dinner, I asked Wendee to join me as my date. Although at first we thought we had little in common, we discovered during the course of our friendship that, for what really counts, we actually had very much in common.

In July 1996 Wendee and I went to Paris for a conference on Shoemaker-Levy 9. Atop the Eiffel Tower on a cloudy, windy evening, I proposed to her. As we looked out over the streets of Paris, I asked Wendee to imagine the city as part of a world with sunrises and sunsets. Then I asked her to close her eyes and imagine another world, a much smaller, more rocky orb that also has its sunlit days and starry nights. "I cannot give you the Moon and stars," I went on, "but I can offer you this small world." When she opened her eyes, Wendee was looking at a plaque announcing that the International Astronomical Union had named asteroid 6485 *Wendeesther* in her honor. "Through her inspired teaching, leadership, and example," it read, "she has touched the lives of thousands of young people."

Upon our return, we started looking for a home. We located one in Vail, Arizona, only 30 miles southeast of Tucson, with a relatively dark, pristine sky. On this two-acre property I designed and built a 12-by-12-foot sliding-roof observatory in the backyard last summer.

Our wedding date did not come by design. We had chosen to marry at the Flandrau Science Center in Tucson last March 16th, but the college basketball tournament that weekend would make it almost impossible to park near the planetarium. So the date was put off a week, to March 23rd. When we realized that the Moon would be in near-total eclipse that evening, we planned a dinner reception under its coppery red gaze.

With so much of my life devoted to astronomy, my observing sessions with Wendee are in a sense a renewal of my astronomical vows. As I show Wendee the night sky, I feel I'm getting acquainted with it all over again.

Some of our best times together have been spent under the stars. Recently we went out to celebrate my 10,000th observing session since I saw a solar eclipse in 1959. Wendee and I quietly walked into our home observatory, sat down, and, like Walt Whitman, looked up in perfect silence at the stars.

SEPTEMBER 1997

**Author Update:**

*This book is coming on the 10th anniversary of our marriage.*

My wife, Wendee Wallach-Levy, and I with the 16-inch Dobsonian at our new backyard observatory in Vail, Arizona. This is the same telescope I used to discover seven comets from my previous home in Tucson.

# Four Decades of Stellafane

Forty-one years ago, Stellafane was unique. The only major amateur event in the world that took place under a dark night sky rather than in a meeting room, this annual summer gathering on a Vermont hilltop was the place to be if you wanted to know the latest in telescope making, observe with old friends, and make new ones.

We live in a different time now. *Sky & Telescope*'s Web site lists no fewer than 40 star parties taking place during the first few months of 1998 alone. But when Constantine Papacosmas first heard about Stellafane as a young amateur in the late 1950s, it was the only show in town, and virtually the world. In the summer of 1957 he drove down with some friends in a car so finicky that he dared not turn off the engine from the time he left Montreal, Canada, until he arrived at the campground many hours later. For Constantine, that first Stellafane offered telescopes, people, and stars on a global scale. It was a taste he would not soon forget.

Constantine Papacosmas in front of Stellafane's famous Pink Clubhouse atop Breezy Hill in Springfield, Vermont. Papacosmas has been regularly attending Stellafane since 1957.

Last year marked his 41st consecutive Stellafane.

Those early meetings were small and intimate. Just about everyone camped in the large, grassy field just east of the main site. Constantine recalls visiting, in his first year, a small cabin in the woods west of the Pink Clubhouse. This was the secret gathering place, a spot few knew about. Providing a magnificent view of the Green Mountain scenery, here was where one could get away from the crowd and take in the Vermont sky.

One time it rained so heavily that more than 300 cars bogged down in the camping area, and a farmer's tractor pulled them out for $3 each. On clear nights the fog sometimes settles over the town of Springfield two miles away, blocking all lights and turning Stellafane into an island in the sky.

Although he didn't get a chance to meet Stellafane's founder, Russell W. Porter, who died in 1947, Constantine did encounter other colorful personalities at Breezy Hill, including Allan Mackintosh of the Maksutov Club; Robert E. Cox, who edited *Sky & Telescope*'s Gleanings for ATM's department; and the young, pipe-smoking Walter Scott Houston, whose evening talks, called Shadowgrams, reminisced about Stellafane's history.

The early 1970s brought about a transformation at Stellafane. When I went there in 1974, the campsite had grown more crowded and the atmosphere less congenial. "Before, everybody could leave their binoculars, cameras, and eyepieces lying around on their blankets," says Constantine. "You'd be sure to find them undisturbed when you returned." Today, as in the rest of the world, people generally do not leave anything unattended.

Things are constantly changing at Stellafane. In 1957, Constantine recalls, "the tradition of Porter was very much evident. People made everything — the optics, the tubes, the mounts — even the screws were handmade." With the rapid growth of telescope manufacturing firms, Stellafane now sees more and more commercial instruments. "There is not as much talk on optics as before," he laments.

Constantine also bemoans the passion of some members of the amateur community in using telescopes to see ever fainter galaxies and other deep-sky objects without any real interest in learning about their true nature. Exploring the night sky ceases to be a science this way; it becomes merely a sport where the one who sees the most and the faintest is the winner. While Constantine means what he says, he doesn't intend this as criticism but as a statement of his own perception.

I first met Constantine in 1960, and within a few months we became good friends. Like many veteran Stellafaners, Constantine sees the annual gathering as a pilgrimage not to be missed. "Stellafane is for creating something yourself by taking a piece of junk and fashioning something out of it that's worthy of the stars," he says. "It's a feeling that goes back to the days of Porter."

Constantine likes the simplicity and unhurried ambiance of Stellafane. And though he has made several telescopes, including the 16-inch Dobsonian I use for my comet hunting, he has no plans of entering a competition. "I don't come to compete; I come to see what others have done and to learn from them."

Just once did I persuade Constantine to break with his tradition. In the spring of 1979, I developed an idea for a "talking" telescope. Constantine, my friend Mikael Stoffregen, and I constructed a 4-inch f/15 folded refractor housed in a working loudspeaker box. We named the instrument Mintaka, after one of Orion's Belt stars.

Collimating its optics proved more difficult than we thought, and Constantine spent most of the drive down to Stellafane tweaking the system. Arriving at noon, we barely got Mintaka entered in time for the competition. I believe it was the first time Stellafane had seen an instrument that combined a 10-inch woofer speaker with an altazimuth-mounted refractor. Throughout that afternoon the British-accented female voice of Mintaka (played from a tape recorder) enlightened visitors on how the telescope worked. At the end of the day we were thrilled to receive third prize for our creation.

On the 41st anniversary of his first Stellafane, Constantine and I sat by the steps of the Pink Clubhouse and talked about what the years had done. The site has changed. The old camping area, which was on land that Stellafane did not own, is now a Christmas-tree farm. But Stellafane still has retained the feeling of old, where a much younger Constantine looked up at a sky so black he could reach out and touch the stars.      JUNE 1998

### Author Update:

*In August 2006, Constantine celebrated his fiftieth consecutive Stellafane.*

# The Man on the Moon

*And, when he shall die,*
*Take him and cut him out in little stars,*
*And he will make the face of heaven so fine*
*That all the world will be in love with night,*
*And pay no worship to the garish sun.*

— William Shakespeare
*Romeo and Juliet*, 1595

There was a hush in the crowd gathered at Cape Canaveral in Florida last January as the final countdown began. Three miles to the east, bright searchlights soared into the heavens as if to point the way for NASA's latest robotic explorer, Lunar Prospector, which was poised atop an Athena II rocket. Locked deep within the spacecraft was a tiny capsule containing one man's greatest dream — to go to the Moon.

"It all came to me in a flash one sunny morning in 1948 while driving to work," Gene Shoemaker recalled. Gene was only 20 years old and living in a mill camp in Colorado for a diamond-drilling project when it dawned

Night turns into day as Lunar Prospector rides on a pillar of fire on its way to the Moon on January 6th. The spacecraft carries onboard some of the cremated remains of Gene Shoemaker *(left)*, fulfilling his life-long dream of going to the Moon.

on him: Someday the United States will send a man to the Moon. Who would be the best person to go? A geologist, of course! At that moment Gene shifted his car into high gear and resolved to be that person.

By 1963 Gene had already begun to define the science of astrogeology that would provide the *raison d'être* for his journey. NASA had also started the process of selecting its future Apollo astronauts. Gene's dream was right on track until that spring, when his adrenal glands began to shut down with the onset of Addison's disease. Although the functions of his glands could be easily performed through hormone therapy, he was no longer medically qualified to go to the Moon. To the end of his life, Gene never forgot his biggest disappointment. "I still dream about walking on the Moon," he would say, "but I had to do other things."

Then came the July 18, 1997, car crash that killed Gene and seriously injured his wife, Carolyn. Sending some of Gene's cremated remains to the Moon was the brainchild of Carolyn Porco, a former student of Gene's at Caltech and now a well-known planetary scientist at the University of Arizona's Lunar and Planetary Laboratory in Tucson. The day after Gene's death, Porco wrote to me about her idea and asked if I would suggest it to the family. On July 20th I was able to reach Carolyn Shoemaker, who was recovering in a remote hospital in the Australian outback. In what was the saddest conversation I had ever had with my dear friend, I mentioned the idea. Although she was a half world away, I felt Carolyn light up at the suggestion. "Tell Carolyn Porco," she said, "that the Shoemaker family would be thrilled if this tribute could happen."

Indeed, things had to happen quickly for this special tribute to make it to Lunar Prospector's launch. Gene's capsule had to be loaded into the spacecraft before it was spin-balanced, a test scheduled early that fall. In one hectic weekend in late August Porco drove from Tucson to Phoenix to oversee the fabrication of the capsule's inscription, then continued to Flagstaff to take part in a small family ceremony at the Shoemaker home during which an ounce of Gene's ashes was selected. She then personally delivered everything to NASA's Ames Research Center near San Francisco, California. Later, mission manager Scott Hubbard coordinated the delicate task of placing the precious cargo into the spacecraft.

Lunar Prospector and Athena were slated for launch on Monday, January 5, 1998, at Cape Canaveral's Pad 46. The Shoemaker family gathered at the Kennedy Space Center; my wife, Wendee, and I were invited to join them. At the prelaunch briefing Wesley Huntress, NASA's Associate Administrator for Space Science, made a moving connection between Lunar Prospector's exploration of our nearest neighbor in space and Gene's legacy. When Huntress concluded, "We're doing this for you, Gene," I could picture Gene, ready to climb aboard, smiling at him from the front row.

A few hours later we assembled at the viewing stands some three miles west of Pad 46. The sky was clear and the Moon was waiting, but a defective ground-tracking radar delayed the mission for 24 hours. We were heartbroken — if the launch didn't take place the following night at 9:27 p.m. Eastern Standard Time, within a four-minute window, the mission would be postponed for a whole month.

So we waited out the next day, January 6th. The Shoemakers, Wendee, and I gathered again early that evening. This time the spacecraft, rocket, and radar were ready, but an approaching thunderstorm threatened to be on top of the Cape at the moment of launch.

"The sky's clearing!" Wendee called out at about T minus 20 minutes. We could now see stars shining in the east, and soon the first-quarter Moon peeked through the clouds. As Launch Control announced the final seconds, everything seemed to stop. The crowd fell silent, the Moon paused in its orbit — everything was still, save for a terse, loudspeaker voice booming, "Five . . . Four . . . Three . . . Two . . . One . . . Zero!" For a split second there was darkness and silence all around. Then the whole eastern horizon suddenly lit up with a brilliant, orange glow as bright as the Sun. Slowly, stately, the Athena rocket broke its ties with the Earth. Climbing higher and faster, the rocket disappeared into the night.

As you read these words, Lunar Prospector is busy studying the geology of the Moon. The mission will end as it crashes onto the Moon's surface sometime before the end of the century. At that moment, Gene Shoemaker will be the man on the Moon, his dream fulfilled at last.       JULY 1998

## Author Update:

*On July 31, 1999, after orbiting the Moon for eighteen months, Lunar Prospector crashed near Luna's south pole. You can visit Carolyn Porco's tribute to Gene Shoemaker at http://ciclops.org/public/tribute.html.*

# An Observer for All Seasons

It's late at night; dawn is not far away, and you've been observing with your telescope for hours. You have that mixed feeling of great accomplishment and overwhelming fatigue. At such a time do you ever imagine that some of the greatest skywatchers of the past are out there, cheering you on? One of them is probably George Van Biesbroeck, a man whose life was spent studying the sky virtually every clear night from 1895 until 1974.

Van Biesbroeck was born in Ghent, Belgium, on January 21, 1880, into a family of artists. An astronomy book he read at age 15 set off a passion for studying double stars, and later comets, asteroids, satellites, eclipses, and variable stars that would span more than three-quarters of a century.

Van Biesbroeck's father understood the financial hardships of being a professional astronomer and he wanted his son to pursue a different career. To please his father the young Van Biesbroeck obtained a degree in civil engineering in 1902 from the University of Ghent. For a few years after graduating Van Biesbroeck was busy constructing roads and bridges. But the call of the sky didn't leave him; he volunteered as a night observer at the Royal Observatory in Uccle, and eventually the observatory offered him a small stipend.

In 1905 Van Biesbroeck accepted a fellowship position at the University of Heidelberg, where he studied under the famous asteroid and comet expert Max Wolf. By 1914 his life had become a nightmare. The invading German army had taken over the observatory, and since Van Biesbroeck could speak German, they ordered him to help guide their troops through the city. Fearing that if he stayed at Uccle he might be sent to a concentration camp, Van Biesbroeck applied for a position at the University of Chicago's Yerkes Observatory to assist Sherburne Wesley Burnham in his work on double stars. To his delight Yerkes accepted him and in 1915 he headed for America.

The double-star project was completed the following year, but on his way back to Belgium he was ordered to report to the Belgian consulate in Utrecht, Holland, to await his enlistment into the army. Back at Yerkes Burnham was retiring, and observatory director Edwin B. Frost promptly invited Van

Biesbroeck to return permanently to Yerkes as Burnham's replacement. With wartime travel to the United States virtually cut off, the Van Biesbroecks had to wait in Utrecht. Finally, in June 1917 they were able to cross the Atlantic to America. The haggard family arrived at Williams Bay, Wisconsin, with only 9 cents in their pocket.

Yerkes was a wonderful place, and it was there that Frost gave the young astronomer his popular nickname, "Van B." At Yerkes Van B developed two areas of expertise — double stars and comets. Double-star observing requires an enormous amount of time and patience at the eyepiece. It is also a field that relied on visual acuity long after photography dominated most other areas of research. (Since long exposures were required for the slow films used at that time, double-star observers could not take advantage of good moments of steady seeing. Thus, the components of a double star would appear to coalesce into one unmeasurable image on photographs.) Van B logged more than 36,000 double-star measurements in his career.

George "Van B" Van Biesbroeck (1880–1974) with the 40-inch refractor at Yerkes Observatory. For more than 70 years the indefatigable Belgian-born astronomer made important observations and discoveries of comets, asteroids, and double stars.

Comets became a highlight of Van B's life on November 17, 1925. He was using the large finderscope of Yerkes' 40-inch refractor to recover the faint comet Orkisz when he spotted a completely new, 8th-magnitude fuzzy visitor. He found a second comet in 1936 and a periodic one in 1954 that rounds the Sun every 12 years. But his comet finds were incidental to the contributions he made to astrometric observations of these bodies. Our understanding of the orbit of almost every comet seen during the middle 50 years of this century is largely due to Van B's work.

On one of the many cloudy nights at Yerkes, Van B and fellow staff member Otto Struve started thinking about building a large reflecting telescope in a place that was dark and always clear. They eventually chose a site in the Davis Mountains of southwestern Texas, near the town of Fort Davis. The 82-inch reflector Van B helped build at McDonald Observatory, as the facility came to be known, provided him with countless nights of observing under some of the clearest skies in the world.

When Van B retired from the University of Chicago in 1945, he continued observing at McDonald and at other observatories. He also developed a new interest: chasing solar eclipses.

In 1963 the Van Biesbroecks visited Tucson, Arizona, en route to a vacation in California. While there they visited Gerard P. Kuiper, a longtime friend who was then director of the University of Arizona's Lunar and Planetary Laboratory (LPL). Kuiper suggested to the 83-year-old Van B: Why not start all over again on a new career?

On August 1st of that year Van B joined the staff of LPL. He continued his visual measurements of double stars and positional observations of comets and asteroids with that institution's 61- and 16-inch reflectors. In 1964 he celebrated his 84th birthday with an observing session at Kitt Peak's 84-inch reflector. Two years later Van B was bemused to see the *IAU Information Bulletin 17*, dedicated, or so it read, to the "Memory of the Late Prof. G. Van Biesbroeck." A letter darted out of Tucson stating that the dedication was somewhat premature. The letter was signed by 36 astronomer-witnesses, including Van B himself.

Van B's passion for observing was unequaled. While discussing how precarious an observer's situation was inside the prime-focus cage atop a telescope, asteroid-hunter Tom Gehrels noted that a fall from that cage would surely be fatal. "What a way to go!" replied Van B. Still going strong in 1970, Van B marked his 90th birthday by observing with Steward Observatory's 90-inch telescope. The following night a misstep caused him to lose his balance and fall off the observing platform. He struck his head on the floor and lost consciousness. When Gehrels visited him in the hospital, Van B seemed barely alive. Unsure what to say to the unconscious man, he gamely mentioned that he had just observed Comet Tago-Sato-Kosaka. Hearing those words, Van B opened his eyes, pointed a finger, and asked Gehrels if he had seen that the comet had split!

That March, while driving from Florida, where they had observed a total eclipse, Van B and comet scientist Steve Larson stopped by McDonald Observatory. Since they arrived just after visiting hours, the attendant at the observatory's visitor center refused to let them in. But Larson persuaded her to call the office. "There's an elderly gentleman here named Van Bies . . . something," she murmured reluctantly. "Claims he used to work here." At the other end of the line, the astronomer's jaw dropped. Send them right up, he insisted. "When we got to the summit," Larson recalls, "people were appearing from everywhere, almost from out of the woodwork. They all wanted to see Van B." As he and Larson left, they rounded a corner that afforded a final view of McDonald. Van B asked to stop. He stepped out of the car and took a long, last look at the place he had built and used for a half-century.

Although George Van Biesbroeck has been gone for a quarter of a century, he is still one of the best examples of a modern Renaissance astronomer. His contributions to positional and dynamical astronomy, as well as observatory construction, were formidable. His zeal for observing was endless — he kept up his rigorous pace, even during subfreezing nights, almost to the time of his death on February 23, 1974, at age 94.

When a *Milwaukee Journal* reporter once asked him what church he went to, Van B waved at the sky and replied, "the Big Church." So when you're outdoors communing with the stars on a dark, clear night, imagine that Van B is at your side. Just one more night under the starlit canopy is something he would have wanted, too.    MARCH 1999

### Author Update:

*Since 1997 the American Astronomical Society has awarded the Van Biesbroeck prize, which honors a living individual for long-term unselfish service to astronomy, often beyond the requirements of his or her paid position.*

# Letter to My Granddaughter

Dearest Summer,

You're the youngest person I've ever written about in Star Trails. Born March 14, 1998, you're 118 years younger than last month's subject, the late renowned astronomer George Van Biesbroeck. As you relax at your parents' home in California, you have no idea what an exciting and invigorating life is in store for you. Your parents, Nanette and Mark, already own a telescope, so somehow, — no pressure now! — I have a feeling that your eyes are going to be exposed to the wonders of the night sky at an early age.

Summer, right now your universe consists of not much more than the rooms of your house. But it is rapidly expanding to include your neighbors' houses, shopping centers, and the Los Angeles basin you call home. Soon you'll understand that you live in a country called the United States. Finally, one bright day you'll realize that your real home is the Earth itself, a spherical body that travels around the Sun, celebrating each year's completed journey on your birthday.

I hope that your first night under the stars will be truly a night of discovery, as it was for me — with no star atlas in hand, I had only to look up at a group of anonymous stars. I pointed my telescope at the brightest one and suddenly realized that it was Jupiter with its entourage of moons. For me that first night was a time of unparalleled discovery. Just me and a group of unnamed stars, all waiting to be arranged into my own little constellations. What star patterns will you invent on your first night out? A puppy dog? A beach ball in the night? Through the stars you can have anything you like.

Sooner or later, you might want to learn the traditional names of the constellations, for they teach you not just about the sky but also about millenniums of human history, with the hopes, dreams, and aspirations of people who lived long ago. You'll also see Wendee's Star, the second brightest star in the Little Dipper — always up in the northern sky; I informally named it for your Grandma when we were courting.

You and your generation will have unprecedented access to more information than you'll know what to do with! Who knows what you'll

learn on the Internet, which is already a huge and convenient storehouse for information. By the time you're ready for it, the Internet will be as easy to use as your television, offering you instant information on what's in the sky each night, when the Moon sets, and where the planets are. That's the biggest change between your life and what I had to do when I was young. But, Summer, don't forget the old-fashioned books. Their pages will tell you so much about your favorite subjects, and when you open a book, it's as though its author, who might have been gone for centuries, has just invited you to travel with him. I hope you'll like each voyage.

Living in the Los Angeles area, the splendor of your night sky will be diminished by the huge amount of unwanted light that the residents pour heavenward. Unless your family drives out of the city, you probably won't be able to see the beautiful Milky Way, the heart of the galaxy we call home. All the sky's dimmer stars are extinguished by that ubiquitous scourge we call "light pollution." Undoubtedly it will get worse during your lifetime. But George Eslinger, Los Angeles's lighting engineer, recently told me that the city's 240,000 streetlights are being redesigned to become more sky friendly. So your city government is working to preserve the night sky for you and everyone else in your generation.

Do you think you'll want to be a scientist by profession? The number of jobs in science, especially astronomy, has been declining, but it seems to me that this decline has slowed down in recent years. If you're good, and I know you will be, you might achieve great success as an astronomer. To do that, you'll want to learn all the mathematics and physics you can get.

I'd be delighted if you just grew up to appreciate the beauty of the sky and the world around you. If you go to a public school, you might get about a month's worth of astronomy in all the years of your schooling. A private school might, or might not, do better than that. A lot depends on your teachers; if they are enthusiastic about nature, they might well be interested in the sky and share with you their enthusiasm. Or you may have a chance to get exposed to one of the astronomy projects that makes it to the school system. In California there's Project ASTRO. Sponsored by the Astronomical Society of the Pacific, it pairs educators and astronomers as teaching partners. If that happens, you'll get to meet a real astronomer besides me! But in all likelihood, the sky is something you, your friends, and your classmates will get to know on your own.

The important thing, Summer, is that you allow yourself to be exposed to the beauty that is all around you — the plants in your garden, your dog's barking, a view of the Earth from an airplane, and the precious darkness of the night sky. Summer, it's all ahead for you; may your life be as rich as there are stars in the sky.                          APRIL 1999

All my love,

*Grandpa David*

### Author Update:

*At the tender age of eight, Summer wrote a story for her class about how she discovered a new planet, with rings, that she called "Bright Spring."*

# Skyward Bound

When an astronomy club goes through a period of intense activity, the credit often goes to just one or two "live wires" whose enthusiasm and organizational talent tend to jump-start the entire group. That's how it was at the Montreal Centre of the Royal Astronomical Society of Canada (RASC) for a period of 30 years beginning in 1940. The person behind this extraordinary era was Isabel K. Williamson, who passed away on June 2nd at 92. During those years Williamson was a mentor to countless new observers, including me (see page 38).

I first became acquainted with Williamson way back on October 8, 1960. It was a Saturday, and for many amateur astronomers living in Montreal that meant a trip to the local observatory, where the RASC was holding its weekly meeting and observing session. It was springtime in astronomy for me. I had just started to use my first telescope, and on this magic night my brother, Richard, and I were about to share my new hob-

Isabel K. Williamson (center) was one of the most active members of the Royal Astronomical Society of Canada's Montreal Centre. For three decades, she helped organize the RASC's observing programs, teaching newcomers how to view the Moon, meteors, auroras, and deep-sky objects. Here she conducts her Lunar Training Program, circa 1960.

by with real astronomers. Isabel Williamson was the first person we met after walking into the observatory. Greeting me with a bright smile, she asked me a few questions and then made a specific suggestion: I should buy *Sky & Telescope*'s Lunar Map, which was available for 25 cents at the time. She said I should try to observe and identify all 326 lunar features plotted on it and then draw them on a fresh map.

That evening helped define the course of my career. It was a call to action. Here I was, at age 12, being asked to do a scientific experiment that was way beyond what I was used to at school. This Lunar Training Program was one of Williamson's ideas to turn armchair astronomers into active observers. Another was the Messier Club. Today, the Astronomical League is but one of several organizations that offer certificates to people who successfully hunt down all 110 deep-sky objects listed in Charles Messier's famous catalog, but the club Williamson started in the early 1940s was the first in North America.

"Its main purpose," she wrote, "was to stimulate members into becoming active observers instead of being content to look through the telescope at objects that others had located." To receive credit, the observer would have to use his or her own telescope and manually sight each target through the main scope or finder by star-hopping. In earlier years, the use of setting circles was prohibited, but that rule was relaxed later on. (I wonder if Williamson would have allowed the use of computerized Go To telescopes.)

Although she didn't graduate from college, Williamson was an avid reader and learner throughout her life. At 16 she gave up a scholarship to McGill University in order to help her family through difficult financial times. Bitten by the astronomy bug, she joined the Montreal Centre in 1942 and quickly mastered the skills necessary for a visual observer. What interested her most, however, were programs that required organizing and training groups, such as meteor observing. She formed teams to monitor summer's major showers, such as the Delta Aquarids and Perseids. They followed the method endorsed by meteor astronomer Peter Millman, wherein the sky was divided into eight sections, each section watched by an observer. During the all-night vigil observers would take turns observing for an hour and then have a half-hour break to avoid fatigue. A central recorder kept a tally of the meteors seen by everyone.

On the evening of October 9, 1946, events conspired to propel a storm of meteoroids toward Earth and Williamson's name into the annals of Canadian astronomy. She had put together a team of 25 observers to keep an eye on the Giacobinid shower, also known as the Draconids, which was expected to be strong that year in the wake of the passage of the stream's parent comet, 21P/Giacobini-Zinner. The team began observing at 9 p.m., and by midnight it had logged nearly 3,000 meteors. Partly as a result of that effort, the RASC presented her with its Chant Medal in January 1949.

In 1948 Williamson put her writing skills to good use by launching *Skyward*, the Montreal Centre's newsletter. I've seen many astronomy club newsletters, some humorous, many filled with technical articles, but *Skyward* was devoted to crediting members who had added another Messier object to their list, separated a difficult double star, or even shoveled snow off the observatory's walkway. Williamson felt that acknowledging the observers' work would motivate them to stay in the hobby. The newsletter's slogan, "You saw it first in *Skyward*," was actually a powerful message to its readers. The publication would contain no reprints of articles easily found in other magazines or newspapers (or on the Internet, if she were still at it today). It did offer original articles of interest to members, as well as reports of the Centre's activities.

From 1956 through 1958 many nations joined the effort to study Earth during the International Geophysical Year (IGY). For American amateurs the IGY was marked by Moonwatch, a worldwide network of visual observers organized by the Smithsonian Astrophysical Observatory to track these new satellites. Montreal, being a bit too far north of the paths of most satellites, seemed left out. The Canadian amateur effort went instead to a detailed study of the aurora borealis, in which Williamson excelled. Observing forms were handed out to all RASC members, and we were encouraged to check the sky several times each night for any signs of auroral displays. Until the evening of July 8, 1966, I had patiently checked off the little "no aurora in sky" box at the bottom of the form exactly 801 times over a two-year period. My patience was finally rewarded, for that night a spectacular display began with shimmering green and red rays. This was followed by rayed arcs and flames that covered most of the sky and lasted all night. The fact that Williamson had spurred me to report all those previous negative observations made my first aurora even grander.

Around 1970 Williamson decided to curtail her activities at the Montreal Centre. After that she devoted her energy to her church projects and other work. Although she did make an occasional appearance at the observatory that now bears her name, I thought that perhaps she had lost interest in the night sky after all those productive years. A few months ago, I called her with the news that the book she had introduced me to decades earlier, Leslie Peltier's *Starlight Nights*, was in print once again. Although she barely had the strength to lift the phone's receiver, she did want a copy. What she didn't tell me, however, was that she had recently presented an hour-long talk to her friends on the special joy of being an amateur astronomer. Williamson's love of the sky, it seemed, never left her after all. NOVEMBER 2000

# Arthur C. Clarke's
# Vision of the Cosmos

Every time you watch satellite television or look at a weather-satellite picture, you're making use of a network of artificial satellites hovering approximately 35,900 kilometers (22,300 miles) above the Earth. At that distance the spacecraft are said to be geostationary — that is, they travel above Earth's equator from west to east at an orbital speed matching that of Earth's rotation, thus appearing to stand still over a single spot on the ground. Geostationary satellites are now routinely used for global telecommunications, navigation, and weather monitoring. Whose brain first thought up this crazy idea of shooting artificial moons into outer space so that our lives on Earth would be easier? It was Arthur C. Clarke, the same genius whose epic story, *2001: A Space Odyssey*, depicted a cosmic view of the dawn of humanity and an optimistic vision of human adventure in space. Space is for dreamers and observers, and Arthur C. Clarke is both.

Born in Somerset, England, in 1917, Clarke discovered the world of science fiction when he was 11. In 1934 he joined the British Interplanetary Society, and two years later he moved to London. During World War II he was a radar officer in the Royal Air Force. He helped run its Ground Controlled Approach radar, the first aircraft blind-landing system, developed by the American scientist Luis Alvarez. (Clarke's novel *Glide Path* is based on this experimental work.) He also spent some time at a radio school not far from the ancient Stonehenge monument.

At the close of the war, the scientists who had built and launched Nazi Germany's V-2 ballistic missiles immigrated to the Soviet Union and the United States. Amid the gypsum desert of White Sands Proving Ground in New Mexico, they worked with their American counterparts to rebuild and relaunch these rockets. The British knew firsthand the destructive power of rockets in war as the V-2s bombarded London; rocket power, as demonstrated by the German V-2s, was becoming a major new force. In 1952 Clyde Tombaugh, the discoverer of Pluto, who worked at that time at White Sands, got Clarke access to the site and talked with him about rocketry's potential.

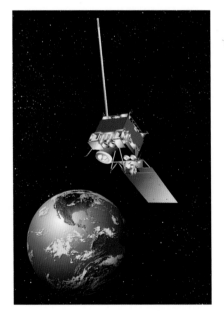

Clarke saw the promise of space travel, but he also understood that governments would be loath to embrace it unless a practical use could be found. In October 1945 *Wireless World* published his article "Extraterrestrial

Relays," which laid down the principle of geostationary satellites. Three communications satellites, positioned 120° apart in a geostationary orbit, or "Clarke orbit," as some now call it, would cover the globe except the polar regions. Clarke's idea was ignored at first, as the U.S. and the Soviet Union struggled for technological superiority in the postwar years. Then on October 4, 1957, the Soviet Union launched Sputnik, the world's first artificial satellite, into orbit around the Earth. American thinking changed overnight. The U.S. was catapulted into the space race, and Clarke's idea for geostationary satellites was finally taken seriously. "Communications and astronautics were inextricably entangled in my mind," he explains, "with results that now seem inevitable." Clarke's vision became a reality in 1969, when the global network of Intelsat III geostationary satellites became operational shortly before Apollo 11's historic lunar landing.

"If I had not proposed the idea of geostationary satellite relays," Clarke wrote me, "half a dozen other people would have quickly done so. I suspect that my disclosure may have advanced the cause of space communications by approximately 15 minutes." Hardly. Clarke, as a science-fiction writer, could see its potential and suggested it first. More important, he wouldn't let the idea drop. In 1947 he wrote *Prelude to Space*, in which he pushed his idea for communications satellites even further. It was a novel set 30 years in the future — he envisioned a world where these satellites played an important role. It turned out to be prophetic indeed. "I have reason to believe," Clarke e-mailed me, "that the proposal had some influence on the men who turned this dream into reality. In the 22 years between the writing of *Prelude* and an actual landing on the Moon, our world changed almost beyond recognition. Back in 1947 I didn't believe a lunar landing would be achieved even by that distant date. I would never have dared to imagine that by 1972 a dozen men would have walked on the Moon, and 27 would have orbited it."

"I believe that if you're an optimist," says Clarke, "you have a chance of creating a self-fulfilling prophecy." In April 1968 *2001: A Space Odyssey* exploded into theaters around the world. Eight months later three real-life astronauts circled the Moon for the first time, sending greetings "to all of you, all of you on the good Earth." The real 2001 seems different from the one Clarke offered us back then. But the differences between what Clarke and the late director Stanley Kubrick foresaw and what we have now are basically in the details. Although they are not yet common, video phones like the ones shown in the movie are certainly available: Clarke used one himself to talk to me during a PBS television program about Comet Shoemaker-Levy 9 nearly seven years ago.

We don't have a base on the Moon yet, and a human mission to Mars is still years away, but a space station is rapidly coming together high above the Earth. Artificial intelligence, cryogenics, and plasma-rocket propulsion are still in their infancy, but we're slowly making progress in these fields. Clarke has no regrets for his futuristic picture of our world three decades ago. In its essence, he had it right.

Thankfully, we still have the benefit of Clarke's insight. At 84 Clarke lives in Sri Lanka (formerly Ceylon), a tropical island nation he moved to in 1956 for its excellent diving, and because he had suffered through "too many English winters." He contemplates the future from a large, book-lined room in a place called Cinnamon Gardens. "A spice odyssey," he teases. "One of my windows looks out on my extensive garden; the other on the blank wall of a ladies' college!"

Daylight has all but faded from my sky tonight. Halfway around the world, Clarke's large garden must now be shrouded in predawn darkness, just as my desert yard will soon be dark. I wonder if he had the chance to use the little Questar telescope he's had since 1956, or his more recent 14-inch Celestron. As I head outside to my backyard observatory, a brilliant moving light grabs my attention. It's the International Space Station which, during the past year, has grown from a faint speck of light to a beacon that can be as bright as Jupiter. Although Clarke has in hand the predicted passage times of the space station over Sri Lanka, clouds often prevent him from sighting it. For me, that distant space station, made real by 2001, is the inspiration of Arthur C. Clarke who, in this real way, has brought space to Earth and made it useful to all humanity.     MAY 2001

# The Street-Corner Astronomer

The corner of Oracle and River Roads in Tucson is one of Arizona's busiest intersections. Each day thousands of cars pass through this rapidly growing northwest section of the city, their drivers hell-bent on crossing the intersection before the next change of the traffic light.

Then they see a 6-inch refractor standing tall in a nearby parking lot, pointed straight at the Sun. Some ignore it, but others slow down, sometimes finding themselves pulling over, parking, and looking through the telescope's hydrogen-alpha filter to gaze at spectacular prominences arching high above the surface of the Sun. One minute the motorists are driving down the street; the next, they're in a different visual universe, staring at some of nature's most ethereal sights.

Starizona is the telescope and astronomy store that offers this free public solar viewing every weekend. It's not the only retail store that does this, but to conduct it at a busy street corner is something special. The idea came from long-time astronomy enthusiast Dean Koenig, whose

Dean Koenig with his 6-inch Astro-Physics refractor. Koenig and his staff at Starizona regularly set up 5- to 7-inch refractors at a busy intersection in northwest Tucson, offering motorists and passersby free views of the Sun, Moon, planets, and deep-sky objects.

goal is to preach the gospel of amateur astronomy to as many people as he can reach. In addition to observing the Sun, Koenig and his staff also host free stargazing sessions four nights a week.

Born on August 3, 1956, at Itazuke Air Force Base in Fukuoka, Japan, Koenig is the son of a fighter pilot who flew missions in Korea and Vietnam. Catching the astronomy bug was, for Koenig, a four-step process. As a child he and his family would travel to the mountains of western Colorado to visit his paternal grandparents. "The place was very rustic," he remembers. One night my father laid me down on the hood of the family's station wagon. As I peered into the unimaginable vastness of the heavens above I was overwhelmed with emotions that still tingle to this day."

Step two involved some help from his enthusiastic younger brother, Kris, whose interest in the night sky began while Dean was a senior at Elk Grove High School in northern California. As president of a junior astronomy club in Sacramento but too young to drive, Kris needed a chauffeur to take him to the club's meetings and star parties. Dean was the ardent volunteer.

Step three was marked by the appearance of Comet West in the pre-dawn skies in 1976. As it rounded the Sun in early March, the comet's nucleus broke into four pieces, resulting in a dramatic surge in brightness. "Kris's 90-millimeter refractor gave the most unforgettable astronomical view I've ever seen through a telescope," he recalls.

After graduating in 1974, Koenig attended Sacramento's Cosumnes River College. There he met his future wife, Donna Gilbert, in a cultural anthropology class. After obtaining his liberal-arts degree in 1977 and becoming engaged to Donna, Koenig moved to Kansas to pursue another powerful interest — his research into the Bible. At The Way College of Emporia he went through the fourth and final step toward his ultimate commitment to astronomy. "That the night sky has biblical significance made the greatest and longest-lasting impression on my life," he explains. "What I learned in Kansas united my passions for biblical and astronomical studies."

Upon his return to California, Dean and Donna got married. They now have three children — Sarah, Kristina, and Steven. "The family grew as fast as my income did," he says with a laugh. "And the purchase of my first telescope kept on getting put off." It was the late 1980s before he and Donna acquired a 6-inch Newtonian reflector as a Christmas present. Bitten by the telescope bug, within a year the couple added a 20-inch Obsession Dobsonian reflector and a 6-inch Astro-Physics refractor. Koenig will never forget those days. "I was hooked," he exclaimed. "Boy, was I hooked!"

For the next 17 years Koenig was in the swimming-pool equipment and supply business. Looking for a change, he contemplated the idea of moving to Arizona and opening a public observatory near Kitt Peak. He visited Arizona several times and trademarked the name Starizona for his proposed facility. "Our family then took the plunge and moved there in 1994," he says.

The enterprise near Kitt Peak didn't go quite as Koenig had planned since the national observatory itself began offering visitors public viewing nights. So his dream was put on the back burner while he set up Starizona as a retail store. "I worked full time for two years as a water-treatment-plant operator while I opened Starizona," he says. "I would get off from work, run home to clean up, and dash off to open the store." Those were difficult years, but in 1996 Comet Hyakutake brought a considerable increase in customers. The appearance of Comet Hale-Bopp the following year kicked the business into high gear.

Not knowing the meaning of a day off, Koenig constantly puts his talent and energy into new projects. I must admit being not totally surprised, for example, when he asked me last August if I'd like to host a radio show. Since I had wanted to do such a show since I was about five, I quickly answered yes! *Let's Talk Stars* debuted only a few days later on the Internet and has been running for an hour a week ever since.

Last fall the International Astronomical Union named asteroid 15321 Donnadean for the Koenigs, but Dean was somewhat embarrassed by the honor. He'd much rather help someone else enjoy his or her telescope or build a backyard observatory. His telescope observations haven't changed his spiritual beliefs; they only serve to reinforce them. "The more we peer at the work of human hands," he says, "the more imperfections it shows. For example, the edge of a sharp razor blade looks like mountains and valleys under a high-powered microscope, yet the work of the Creator shows only more and more perfection when explored through a telescope. There are many things I do not know about astronomy, but one thing I know is for certain — that I'll continue to enjoy with utmost satisfaction my views of the Maker's handiwork."

# The Comet Master

On the evening of February 1st, just four days after full Moon, 58-year-old Japanese comet hunter Kaoru Ikeya brought out his homemade 10-inch (25-centimeter) Newtonian telescope. Just as he has always done on clear nights, Ikeya set the scope on its altazimuth mount in his front yard in Mori, Shizuoka Prefecture, about 195 kilometers (120 miles) southwest of Tokyo. Inserting a 40-millimeter Pentax eyepiece into the focuser, which gave him a 1.6°-wide field at 39×, he began to search for comets by systematically sweeping the sky toward the southwest. "The evening twilight was not completely over yet," he recalls.

Some 30 minutes later, at 6:48 p.m., Ikeya came upon a faint smudge of light in Cetus about 2 arcminutes across with a weak central condensation. "I estimated its brightness to be about 9th magnitude," he says. "The object was very difficult to see visually because of the significant light pollution from the city of Hamamatsu. I carefully identified the object's position and marked it directly on the *Uranometria 2000.0* star atlas." Suspecting it to be a comet, he observed the object over the next half hour. After confirming that it was indeed moving slowly northeastward, he then checked to see if there were any known comets near its location. There were none. Ikeya excitedly notified comet expert Syuichi Nakano, who in turn relayed the news to the International Astronomical Union's Central Bureau for Astronomical Telegrams (CBAT) in Cambridge, Massachusetts.

Meanwhile, 2,200 km to the west near the city of Kaifeng in China's central Henan Province, amateur astronomer Daqing Zhang was also scanning the skies with an 8-inch reflector around the same time as Ikeya. He too spotted the new visitor and promptly reported his find.

Within hours CBAT was announcing the discovery to the world. Comet Ikeya-Zhang, also designated C/2002 C1, soon delighted skygazers worldwide, attaining naked-eye prominence by late February and sporting a slender tail up to 6° long. The finest comet to grace the northern skies since Hale-Bopp in the spring of 1997, C/2002 C1 reached a peak brightness of magnitude 3 in late March. Orbital calculations showed that the object is likely the return of a comet last seen in 1661.

When Ikeya discovered his first comet on January 2, 1963, at age 19,

with a homemade 8-inch reflector, his modest, self-effacing story captured the hearts of many amateur astronomers. By the time a second
Comet Ikeya made its way around the Sun, the name, if not the person,
was already well known to avid skywatchers. But it was his third find, the
great Sungrazing comet Ikeya-Seki of 1965, that made the young Japanese amateur's name a household word. Articles in the press around the
world praised his tenacity and perseverance, and told glowing stories of
how and why he chose to pursue the sport of comet hunting. I remember
being particularly moved by one story published in a magazine that told
how Ikeya's father had failed in his family business, and that his son had
chosen to "take the failed family name and scrawl it across the sky."

Touching as this story was, it isn't true. Born in Nagoya City, Aichi Prefecture, in 1943, Ikeya moved with his family to Bentenjima, near Lake
Hamanakako in Shizuoka Prefecture, and attended elementary and middle
schools in Maisaka. Here he developed an interest in astronomy at age 13.
"My father did own a small business, which failed when I was 15," Ikeya

*Above:* Comet Ikeya-Zhang reached perihelion,
the point in its orbit closest to the Sun, on
March 18, 2002. Japanese astrophotographer
Shigemi Numazawa obtained this tricolor
view of the comet 2° from M31, the Andromeda Galaxy, on April 6th. *Right:* Japanese
comet hunter Kaoru Ikeya with the homebuilt
10-inch (25-centimeter) f/6.2 Newtonian
reflector he used to find Comet Ikeya-Zhang,
C/2002 C1. Ikeya also discovered or codiscovered five other comets, including the great
sungrazer Ikeya-Seki in 1965.

explains. But contrary to press reports, "he did not go to the bar frequently in order to forget his business failure. My family was rather ordinary and our family name was not associated with any special family. There was no family name to be redeemed before and after my father's business failed, so I never decided to try to redeem it by discovering a comet. I've never thought of or said in any similar quote that I wanted to take my family name and scrawl it across the sky."

So why did Ikeya pursue comet hunting? His true story is even more moving than the published fable because it focuses not on his personal ambition but on his reaction to the majesty of the night sky itself. "I love to observe not only the Moon and planets at high magnifications but also star clusters and galaxies at low powers with my homemade telescope," he says. "This is the reason why I started comet hunting. I thought that I could contribute a little bit to astronomy if I found a comet."

In 1962 he began his search. In those days, the sky above Ikeya's observing site near Lake Hamanakako was clear and relatively free from light pollution. "I'm shy and not good at being in public and socializing," he admits. "I seldom go out." The peace and serenity of a dark, star-filled night are ideal companions for a somewhat reserved person, but the pre-dawn hours of September 18, 1965, changed all that. Sweeping through Hydra with a 6-inch scope, he picked up an 8th-magnitude tailless glow west of Alpha Hydrae. Fifteen minutes later Tsutomu Seki, searching independently from Kochi, 400 km away, also spotted the same object.

Some 10 days later and a half world away, Brian G. Marsden was beginning his first day on the job at CBAT. His initial task: to refine the orbital computations for the new comet — Ikeya-Seki, C/1965 S1. With just three weeks left till *perihelion* (the comet's closest approach to the Sun), on October 1st Marsden joined a press conference to announce the coming of the brightest and best-placed Sungrazing comet since 1882.

"I remember it was one to two weeks after discovery that I knew how close Comet Ikeya-Seki would approach to the Sun," Ikeya recalls. "There was speculation that the comet would evaporate and disappear during closest approach."

As Ikeya-Seki neared the Sun, observers could catch a glimpse of it just by covering the Sun with their hands. They estimated the comet's brightness to be magnitude –10 or –11 (about as bright as a gibbous Moon). On October 21st Ikeya-Seki passed a mere 450,000 km above the

Sun's photosphere. In Japan, where the comet reached perihelion around local noon, astronomers at Mount Norikura solar observatory saw the comet's nucleus split into three.

"After perihelion I knew that Ikeya-Seki had survived the Sun's heat when I saw the comet's tail on the other side of the Sun in late October," says Ikeya. As the comet began to fade, its tail became its most spectacular feature, with estimates of its length as great as $60°$.

I'll never forget my first view of Ikeya-Seki as its majestic tail rose over the St. Lawrence River near Montreal. Trying to avoid the poor eastern horizon from our home, I rode my bicycle for more than a kilometer in the predawn cold until I reached the summit of a nearby hill. The comet pierced the sky through the bright lights of downtown Montreal, its tail shining upward like a searchlight beam. It was an indescribable thrill!

Not resting on his laurels, Ikeya continued his comet search. In 1966 he shared a find with American astronomer Edgar Everhart, and in the closing days of 1967 he and Seki codiscovered their second comet. That was Ikeya's last find for the next 34 years.

"I've been continuously searching for new comets since 1967," he writes. "Sometimes I slowed down the search, but I've never stopped my comet hunting completely." By the 1980s his sky was becoming light polluted and less favorable for comet searches, especially near the horizon. Since it was still relatively dark close to the zenith, he concentrated his search to that area. As part of his ongoing project, he carefully checks the structure of each galaxy he encounters during his regular sweeps. In this serendipitous way, in December 1984, using a 10-inch reflector, he discovered his first supernova, SN 1984R, in NGC 3675; four years later he found his second, SN 1988A, this time in M58.

"I will continue to observe and hunt for comets the same way I've been doing in the past — at my own relaxed pace," he says. It's a philosophy that follows the advice given him by the late Minoru Honda more than 40 years ago. Honda, who discovered 12 comets and 12 novae during his lifetime, was responding to Ikeya's letter, written before the latter's first comet find: "If you desperately want to find a new comet, please stop your search because you may never be able to find a new comet. However, if you are content to search the sky without ever experiencing a new comet discovery, please keep searching, because someday you may be able to find a new one."

# Tsutomu Seki and the Great Comet of 1965

On the night of November 4, 1965, I set my alarm clock in the hope of getting a predawn view of the newly discovered comet, Ikeya-Seki (C/1965 S1). Clouds had spoiled several of my attempts earlier that week, but this time the morning sky cleared up and my friends, the Jorgensen family, and I finally got a good look at Ikeya-Seki.

Halfway around the world, 35-year-old Japanese comet hunter Tsutomu Seki was just viewing a picture of the comet he had taken from his discovery site. For many comet lovers, the visitor that Seki helped discover would go down as one of history's greatest comets. For me that morning, it was the first one I'd ever seen, and it was pure magic — not only because of the way its long, majestic tail soared high into the dawn sky, but also because it was first spotted by two amateur skywatchers whose dedication and perseverance helped inspire me to begin my own comet search.

The discovery of Comet Ikeya-Seki on September 18, 1965, was truly fortunate since the day before, a very strong typhoon had passed through Kochi, where Seki and his wife, Okiko, live. "I was half asleep, hearing the sound of roaring winds," Seki recalls of that fateful night. "I didn't know how much time had passed, but then I heard Okiko whispering in my ear, 'Honey, the sky's clear now.' I sprang out of bed and looked out the upstairs window to the north. I saw an awesome starry sky in the wake of the typhoon."

Using his homemade 9-centimeter (3½-inch) refractor with 19× magnification and a 3½°-wide field, Seki quickly began to scan the eastern sky systematically from an observing deck he'd built atop his home. Sweeping through Hydra, he spotted an 8th-magnitude tailless glow west of Alpha Hydrae at around 4:15 that morning. Fifteen minutes earlier, 21-year-old fellow comet hunter Kaoru Ikeya, searching independently from Bentenjima, 400 kilometers away, also spotted the same object. Reports of their findings quickly reached the International Astronomical Union's Central Bureau for Astronomical Telegrams (CBAT) in Cambridge, Massachusetts, the world's clearing-house for astronomical discoveries.

Brian G. Marsden, then with the Smithsonian Astrophysical Observatory, calculated an orbit for the new object and announced at a press conference that Ikeya-Seki was a Kreutz sungrazing comet and could become visible in the daytime. "One week after the discovery," remembers Seki, " I read an article in the local *Kochi Shinbun* newspaper headlined 'Comet Ikeya-Seki, the greatest comet of this century?' I couldn't figure out what it was all about. It was very difficult to imagine that this faint comet would become a great comet."

Born November 3, 1930, in Kamimachi, Kochi City, Tsutomu Seki has had a lifelong interest in music, particularly classical guitar. When renowned guitarist Andrés Segovia visited Japan in 1969, Seki attended one of his recitals and was inspired by his music to master the instrument. Seki has taught classical guitar since then and still offers lessons three times a week.

In 1948 legendary Japanese amateur astronomer Minoru Honda co-discovered the faint periodic comet, Honda-Mrkos-Pajdusakova. Seki was fascinated by Honda's discovery and began to study astronomy on his own. The idea that "a comet bearing [his] name might travel the universe" intrigued him, and, on the advice of Honda himself, the following year Seki ground a 4-inch mirror for his first telescope.

"On August 8, 1950, I started my comet search using this telescope," says Seki. "There were many Perseid meteors that night and I searched the eastern sky until dawn to my heart's content." That first night was a harbinger of quiet, dark nights to come; all six of his comets were found from his rooftop site. But before Seki could claim a comet of his own, he had to endure 11 years of searching —11 years of patiently learning the

Japanese comet hunter Tsutomu Seki with the 3$\frac{1}{2}$-inch refractor he used to discover three of his six comets: Seki, C/1961 T1; Seki-Lines, C/1962 C1; and Ikeya-Seki, C/1965 S1 (pictured to his right). In addition to comets, Seki found 222 asteroids photographically at Geisei Observatory in Kochi.

sky, 11 years spent wondering if the next field would yield a new comet.

At last, on October 10, 1961, Seki discovered an 8th-magnitude object in the dawn sky using the 3½-inch refractor. "I wasn't sure if it was a new comet. I rode my bike to a telegraph office two kilometers away. I was so excited I couldn't think of anything else while pushing the pedals. Mr. Honda verified my discovery the following day. A telegram was sent to Copenhagen [where CBAT was then located] and Comet Seki was announced."

On February 4, 1962, Seki discovered another comet that rapidly reached naked-eye brightness. Comet Seki-Lines, codiscovered by Richard and Helen Lines from Phoenix, Arizona, attained magnitude –2.5 in early April. This was followed by four other visual finds, including a second Comet Ikeya-Seki at the end of 1967. "I felt the presence of the young comet hunter Mr. Ikeya," Seki recalls. Only 15 minutes separated Ikeya's sighting from that of Seki back in 1965; this time the difference was only five minutes. "I believe I was able to continue comet hunting because of good competitors like Mr. Ikeya," he adds. Seki still has a 4-inch mirror that Ikeya ground for him "as a symbol of our friendship."

Although it has been a long while since Seki's last discovery, he still searches for comets. He maintains a Web site and conducts astrometric observations of comets and asteroids with Geisei Observatory's 24-inch telescope. "I live for astronomy," says Seki. "As long as I can hope to find new objects, I will keep observing for the rest of my life." JANUARY 2006

places

# The Real Unified Theory

How many of us started out, eyes wide open to the stars, wanting to be a professional astronomer, at the controls of some of the world's most powerful telescopes? Chances are, if you've been a long-time reader of *Sky & Telescope*, that dream has crossed your mind at some point. But odds are also that your career ended up taking a radically different path.

Many of us specialized too early, partly due to a perceived need to get a good start with life — enroll in a good university and move into some narrow field directly toward a degree and a good-paying job. In our practical world, that's probably good advice. But it means that most of us can never follow our dreams. Not so in Southern Illinois University at Carbondale's University Honors Program. Thanks largely to its director, Frederick ("Rick") Williams, the program brings together students with varied interests and gives them a chance to pursue their dreams.

Rick Williams is not an astronomer in the classical sense. Born in Denver, Colorado, in 1942, he grew up in the small town of Foxpark, Wyoming. His grandfather, a schoolteacher, planted some basic astronomical knowledge in the inquisitive youngster. "He showed me that two of the stars in the Big Dipper's Bowl always point to the North Star," Williams recalls. After graduating from high school he attended the University of Wyoming for two years, where his first interest was prelaw. Uncertain of his own career, Williams left school after his sophomore year. He worked for a while stringing power lines to the Minuteman missile silos that were springing up at many sites around the country at the time.

A year later, having made enough money so he could fund his education, Williams enrolled at the University of Texas at Austin. In order to graduate, he would need to take an additional year of mathematics. But the school offered an unusual choice — students who didn't want the extra math could take a year of classical Greek. "In my junior year I signed up for Greek," says Williams, "and after six weeks I realized I was in love with it." After graduating, Williams went on to the University of Colorado at Boulder for his master's degree and then to Cornell University in Ithaca, New York, where he earned a Ph.D. in classics in 1975. Two years later he began his career at Southern Illinois University (SIU).

With his eclectic background, Williams was the ideal choice to head the university's honors program. He became its director in 1987, the same year he and his fiancée, Brigitte, an architect, were married. "It is the aim of the honors programs," he said, "to try to shape the student not just in one academic subject." But many universities design their programs as a sort of shadow curriculum, so that a physics department offers a standard course in physics, for example, and a more advanced course in the honors program. SIU's program is different. Any student with high academic grades is welcome to join. "Membership is a pat on the back for a job well done," Williams explains, "not a set of obstacles to be overcome."

Over the last 12 years the honors program has developed 45 new courses. An English course, called Art and Science in the Romantic Period, explores a heightened interest in humanity's place in the world and the universe. Philosophers like Pierre Simon de Laplace were investigating new explanations of Nature. Ideas like Laplace's nebular hypothesis, put forth in his *Exposition du système du monde* — an attempt to understand how the Earth and its neighboring planets were formed — were prevalent at the time. Thus, a course like this brings together science, literature, and history.

The honors program has other benefits, one of which is a series of breakfast seminars with visiting lecturers. It was at one such seminar

Rick Williams and I (back row, third and fourth from left, respectively) relax with Southern Illinois University students and faculty after an all-night Leonid observing session in 1996.

that I first met a sampling of SIU's 2,000 honors students — students who were learning the classics, modern languages, English literature, and physics. The session was just between me and the students; no faculty members were there to moderate or grade the discussion. Our interaction lasted more than an hour, covering both the humanities and the sciences. We talked about how Lucius Annaeus Seneca found time to write one of the most important classical treatises on comets ever composed when he was facing execution at the hands of the Roman emperor Nero. Another student asked how comet impacts could have affected the origin of life on Earth. When I explained how comets deposited water and a rich assortment of organic compounds when they collided with our planet, a third student made the connection between chemistry and the evolution of amino acids and DNA. During the session the students built a unifying academic force out of the various branches of the arts and sciences. I left the session invigorated.

How many of SIU's honors students will eventually become professional astronomers? Probably very few. But thanks to this unique academic program, students, faculty, and friends were subjected to a direct lesson from the night sky. On the evening of November 16, 1996, more than a hundred people gathered at the Touch of Nature park to observe the Leonid meteors, then beginning to gather strength in anticipation of the parent comet's return. The mostly overcast skies didn't prevent us from sighting a few meteors, including a bolide that lit up the clouds. But the night was an opportunity for much more than counting meteors. There were discussions about comets, impacts, and history.

By dawn the next morning, those who survived the night and solved the problems of science and the world were treated to a congratulatory breakfast. Not all those students will get their hands on the world's largest telescopes, but they did get a taste of what a university could provide. A university education is like a window to the universe. Looking out through this window they see the many aspects of the world we live in and the beauty of the universe that surrounds it.                    SEPTEMBER 2000


### Author Update:

*Rick Williams now co-chairs the committee for the doctoral thesis that I'm currently writing.*

# A Central American Heaven

Last February, as our plane swung toward the west in preparation for landing at El Salvador's international airport in San Salvador, the thin crescent Moon and Mercury suddenly swung into view through our cabin windows, hanging majestically in the evening twilight. I had never seen Mercury so high in the sky before — a good omen, I thought, for the start of our visit to this land.

Most people think of El Salvador as a nation wracked by civil war. But the war ended in 1991, and in place of a war-torn landscape is a beautiful tropical country with a proud, ancient heritage, a modern, bustling capital city, and some of the darkest, clearest skies I've ever seen. Comet discoverer Carolyn Shoemaker, S&T contributing editor Stephen James O'Meara, and I spent almost a week there to participate in the first International Astronomy Convention organized by the Asociación Salvadoreña de Astronomía (ASTRO) and to inaugurate the group's new observatory.

The four-day convention featured talks at schools, media presentations, and sightseeing tours that culminated in a packed public lecture at the Radisson Plaza hotel. Carolyn talked about the comet-search program that her late husband, Gene, established, Stephen talked about the five greatest revolutions in astronomy over the last millennium, and I introduced the crowd to the passion of observing the stars.

The trip also gave me a chance to get acquainted with Ana Guillermina Reyes, who is one of ASTRO's most active members. A lifelong devotee of the stars, Reyes still recalls her earliest views of the night sky. "I remember being very young, lying on the grass next to our beach house with my brother and parents, looking at the sky," she says. "I remember my mom pointing out to me the groupings of stars that her father had taught her." Orion's Belt, she found out, was known locally as the Three Marys, while the Sword below it was called the Three Wise Men. That early beachside introduction to the stars was further enhanced when her dad set up a small refractor. "How incredible that made me feel," Reyes says of her first look through the telescope. "The wonder of it!"

As in many developing countries, schools in El Salvador lacked any formal course in astronomy. So Reyes and her family invented what she couldn't learn from school, making up simple constellation patterns and names such as the Ladder and the Duck that guided her through her teen years.

In March 1978 Reyes, by this time a young mother, moved with her daughter, Dede, to Houston, Texas, where they bought a small department-store refractor. Their initial disappointment and frustration with that scope prompted them to move up to something better, an 8-inch Dobsonian reflector. The 8-inch led the mother-and-daughter team in early 1990 to join the Houston Astronomical Society. "At first I went to the meeting by myself," Reyes says, "and saw all these strangers, 95 percent of whom were men. I felt a bit out of place then, but I really enjoyed the talks." The meetings led, in turn, to a visit to the Texas Star Party, where she was introduced to big telescopes, Messier objects, star atlases, and her own sense of adventure. "I'll never forget the feeling of triumph at finding on my own my very first Messier object," she remembers. By the end of TSP Reyes had located half the Messier objects, and soon afterward she became the proud owner of a 20-inch Dobsonian.

In 1993 Dede died from complications of a heart condition. A few months later Reyes returned to El Salvador. As she searched for old friends she was surprised to learn that in the late 1980s ASTRO had been formed. Its biggest project at the time was raising funds to build an observatory to house a 12-inch Tinsley Cassegrain reflector donated to the group. Reyes joined in the effort. After several years of hard work and through donations of money and land, they were able to establish a dark-sky site in the town of San Juan Talpa, 40 kilometers from San Salvador. "It was really, really hard," Reyes remembers. "We often ran out of funds during the observatory's construction." By the end of 1997 the facility was finally ready for initial testing.

On February 8th of this year, the dream of El Salvador's amateur astronomers became reality when Prudencio Llach Astronomical Observatory was officially dedicated. Named after a benefactor, the facility features a 5-meter dome, a large lecture hall, and a library. It is open not only to ASTRO members for their observing projects, such as monitoring variable stars, but also to people from all walks of life.

Although El Salvador experiences a very long rainy season, the months of December, January, and February are generally clear and dry. The whole week of our visit we were blessed with beautiful, cloud-free skies. On February 9th some 25 club members and friends from neighboring countries joined us at the observatory for an all-night star party. At latitude 13.5° north, I could see the far-southern Milky Way down to about declination –70°. Using Reyes's 20-inch Dobsonian, I spent hours comet hunting in what was for me virgin territory. I also explored the extremely rich star fields around Eta Carinae, the Southern Cross, and Alpha and Beta Centauri. For those several hours, El Salvador was heaven.      AUGUST 2000

# Crescent City Astronomers

The Crescent City. The Big Easy. These are just two of the colorful monikers that describe one of the most fascinating cities I've ever been to —New Orleans. Bounded on one side by a crescent-shaped meander of the Mississippi River and on the other by Lake Pontchartrain, this largest city in Louisiana takes the definition of character to a new level. After all, what other metropolis would completely close down its streets in order to host a giant Mardi Gras parade with "The Constellations" as its theme? In February 1977 I personally enjoyed that parade, with giant floats representing the likes of Cygnus, Taurus, and Lyra, that wound its way through town.

New Orleans lives in style, and one of its astronomy clubs, the Pontchartrain Astronomy Society, or PAS (www.pasnola.org), is fully in tune with that style. One of the most dynamic amateur groups in the country, the PAS owes its success to its many active members. This month's Star Trails highlights two of its live wires: Michael Sandras and Barry Simon.

Long-time amateur astronomers Michael Sandras (left) and Barry Simon at the Daily Living Science Center's observatory in Kenner, Louisiana.

As a youngster, Sandras traversed the far reaches of the universe at warp speed with *Star Trek*'s Captain Kirk and Mr. Spock. "This original TV series," he says, "made me extremely interested in space travel and astronomy." Visits to the local library, as well as supportive parents who bought him his first telescope — a 60-millimeter refractor — furthered his interest. On the first clear night after he unwrapped his new scope, he aimed it at Saturn, took one look, and was hooked.

After completing high school in 1981 Sandras enrolled at the University of New Orleans (UNO), but he left before graduating to pursue a planetarium career. "Because I didn't finish college," he admits, "I've had to work very hard to prove myself." He started out as an apprentice at Ep Roberts Instruments, a local telescope dealer. In 1989 he was hired as a telescope operator at UNO's observatory. It wasn't long before Sandras was offered a second job, as director of the Daily Living Science Center's planetarium and observatory in the New Orleans suburb of Kenner. There he gives lectures to an average of 20,000 visitors a year. In addition, he is the president-elect of the Southeastern Planetarium Association.

A close friend of Sandras, Barry Simon is a hospital specialist with Novartis Pharmaceuticals. Simon spends his days visiting teaching hospitals in the New Orleans and Baton Rouge areas. His nights and weekends, however, are spent indulging in his passion for the night sky, which began in 1961 when his father gave him a 60-mm refractor. Like Sandras, it was his first view of Saturn that inspired him. "What an amazing sight!" Simon recalls of that night.

In 1975 Simon completed his degree in secondary education at UNO. However, during his college years he let his astronomical interest slide, selling the small reflector he had at the time. It took a movie and a meteor shower in 1977 to bring him back into the fold. *Star Wars* made him think about the stars once again. That same year he had a chance to travel to Panama City in Florida, where late one summer night he found himself on a pier facing the Gulf of Mexico. "It was a dark, clear night, and the Perseids were spectacular!" he exclaimed. Simon's life seemed to come together at that time: he purchased $7 \times 50$ binoculars, got engaged to his girlfriend, Susan, and bought a used 3-inch f/16 refractor. That telescope was the first of a long line of at least 30 refractors that he has either built or bought. Simon particularly loves these instruments, and he currently owns a number of Unitrons, Brandons, and Takahashis.

Simon joined the PAS in 1979. Like Sandras, he has served six terms as club president. In 1983 Simon founded a star party, the Deep South Regional Stargaze, held each fall at Percy Quin State Park in southern Mississippi.

Sandras and Simon have formed an extraordinary leadership dynasty that has given the PAS not only stability but also an almost fourfold increase in membership in the last 20 years. The 200-strong group now has an observatory equipped with an 11-inch Celestron telescope, located some 95 kilometers north of New Orleans.

The PAS offers residents of the Crescent City a wonderful way to enjoy the stars. "We find astronomy to be a very humbling experience," says Simon. "In all its majesty, the sky tends to put everyday problems in perspective. Many people draw inspiration from the night sky when they channel their view through the heart first; viewing the sky certainly does this to us."                                              SEPTEMBER 2001

### Author Update:

*In August 2005, Hurricane Katrina turned much of New Orleans into a ruin. In the year that followed, Barry, Mike, and the other members of the PAS struggled to rebuild their lives. I visited the city in March 2006; parts of it remained utterly devastated. But it was encouraging to discover that through all the destruction, the PAS folks were still taking time to look at the stars.*

# A Peak Experience

There's something magical about spending a night at the national observatory atop Kitt Peak in southern Arizona. The 6,875-foot-high mountain's excellent observing conditions have made it home to nearly two dozen telescopes of up to 158 inches (4 meters) in aperture. I remember my first visit there in April 1963; at 14, I was already an avid amateur astronomer who asked the tour guide, "What does an astronomer have to do to get observing time on one of the telescopes here?" Looking at me coldly, the guide answered, "The telescopes are available only for *qualified* personnel." I spent the rest of that week learning that "qualified" meant not only being a graduate student or having a doctorate in astrophysics but also having an observing proposal good enough to be accepted by the observatory's telescope-time allocation committee.

That was 40 years ago. Despite increasing light pollution from Tucson and Phoenix, Kitt Peak is still a very good site. And a guide today would give me a very different answer to my question. Kitt Peak now allows the public, by reservation, to observe with 16- and 20-inch telescopes every clear night. And there's the Advanced Observing Program (AOP), available also by reservation, for amateur astro-imagers who wish to take CCD images of their favorite objects. Available to all citizens, Kitt Peak is now truly a national observatory.

And so it was that Wendee Wallach-Levy, one of those citizens, was assigned May 22, 2003, for an all-night imaging session with the observatory's 20-inch Ritchey-Chrétien reflector. (She brought along her observing assistant — that was me, her husband.) Adam Block, one of the AOP's guides and telescope operators, walked her through the basics of CCD imaging. As operator, Block was responsible for aiming, focusing, and tracking the telescope at the objects that Wendee wanted to photograph and for processing the images.

We spent the night imaging three objects. The first one was Tombaugh's variable star, TV Corvi, an 18th-magnitude cataclysmic variable that was slowly fading after a recent outburst (see page 104). The second one was a 14th-magnitude periodic comet, 53P/Van Biesbroeck, dis-

The spiral galaxy NGC 6946 in Cepheus as imaged by Wendee Wallach-Levy and Adam Block during a night at Kitt Peak's Advanced Observing Program (AOP). North is to the upper right.

covered in 1954 at Yerkes Observatory by George Van Biesbroeck (see page 60). The third and final object we imaged that evening was NGC 6946, a 9th-magnitude spiral galaxy lying 18 million light-years away on the Cepheus-Cygnus border. It was found by William Herschel in 1798.

Once we were set up on NGC 6946, we made the equivalent of a 90-minute unfiltered exposure with the SBIG ST-10XME CCD camera as well as separate 20-minute exposures through red, green, and blue filters. In the raw images the rich detail of the galaxy's structure began to emerge, but Wendee and I had no idea what excellent seeing, a fine telescope, and the wonder of image processing could deliver. After the individual exposures had been calibrated, Block digitally aligned, stacked, and processed them using *MaxIm DL, MIRA,* and *CCDSharp* software. After applying finishing touches with *Adobe Photoshop,* he displayed the final color composite on the computer screen. There we saw a magnificent "island universe" (to use a term popular a century ago to describe galaxies). I couldn't believe my eyes. To get an image like that with a tiny silicon chip coupled to a 20-inch — a *small* scope by Kitt Peak standards — is simply amazing. It just demonstrates the power of CCD technology.

"Guests of the program are treated just like any visiting professional astronomer," explains Block. "They observe with a telescope operator, they use the same accommodations, have the same meals, and so forth, at no less than the Kitt Peak National Observatory. What better way of showing people what astronomy is all about than letting them become astronomers themselves?"

Summer nights go by quickly, and before we knew it the first sign of dawn tinged the northeastern sky. As Wendee and I prepared to walk back to the dormitory room assigned to us, we saw a brilliant satellite cross the brightening sky. I turned my 3.5-inch Questar telescope on it, and though it was racing through the field of view, I could clearly make out its elongated shape. Sure enough, we were watching the International Space Station pass by. "It was truly a night to remember!" says Wendee.                                          OCTOBER 2003

### Author Update:

*For more information about Kitt Peak's Advanced Observing Program, go to their web site at www.noao.edu/outreach/aop.*

# Miracle at Birr Castle

With a muffled creak, the huge wooden gates swung open, inviting the attendees to the 2003 Whirlpool Star Party onto the spacious grounds of Birr Castle in central Ireland. It was around 11 in the evening on September 27th. As we walked forward, the bright glare of streetlights was replaced by utter darkness. Until our eyes could adapt, we walked into an inky void. The unpaved road curved to the left. Above us the Milky Way shone down placidly, and even M31 beckoned through some thin clouds.

Then suddenly a dark outline loomed ahead, like some lonely iceberg jutting out of the sea — it was a humongous stone wall topped by a wooden arch soaring high into the night sky. Right next to it was a second wall. And as my eyes adapted to the darkness, I could make out a cable beginning north of the walls, going between them, and finally coming to rest at the end of a huge wooden tube. Here it was: the 1.8-meter (72-inch) telescope designed, built, and used by William Parsons (1800–1867), the third Earl of Rosse. And on this magical night we were about to observe the heavens

The 72-inch Leviathan telescope was built in 1845 by William Parsons, the third Earl of Rosse, at Birr Castle in Ireland. Its 58-foot-long tube is lifted by a counterweight system suspended between two massive stone walls. Observers can view celestial objects as they pass near the sky's meridian by climbing the wooden gantry at left.

right in the shadow of this great instrument.

Anyone who knows anything about the history of astronomy is familiar with Lord Rosse's telescope, also known as the "Leviathan of Parsonstown." It was the largest in the world from 1845 to 1917. It's one thing to read about the Leviathan but quite another to confront this monster in person. I first encountered it in the pages of an astronomy textbook when I was 12 years old. The telescope is best remembered for making possible Lord Rosse's sketch of M51, the Whirlpool Galaxy in Canes Venatici. The Whirlpool was the first of more than a dozen galaxies whose spiral structure was confirmed with the Leviathan.

This telescope had an influence on generations of skywatchers, partly thanks to Camille Flammarion's inclusion of the M51 sketch on the frontispiece of *Les Étoiles*. One of those who might have been inspired by that sketch was Dutch painter Vincent van Gogh, whose famous *Starry Night*, painted at the end of June 1889, could have made use of the Whirlpool's vast and gentle swirls of starlight. When compared with Lord Rosse's drawing, the swirls of *Starry Night* seem to match better if the painting is mirror reversed — the spiral motion in M51 seems to be counterclockwise, though the artist made it appear to be clockwise.

The Leviathan showcases a 19th-century Irish nobleman's engineering and ingenuity. Take, for example, its primary mirror. At 72 inches, the mirror was so much bigger than the previous record holder: William Herschel's 48-inch. Lord Rosse had to build a brick oven large enough to handle his four-ton speculum-metal mirror so its molten mass could cool down gradually. He also designed the furnace that would control the oven's temperature. This annealing process is still being used today for making large mirrors, from Palomar's 200-inch Hale reflector to the 8-meter Gemini twins in Hawaii and Chile.

When the Whirlpool Star Party began some 17 years ago, longtime attendees told me, the telescope was essentially abandoned, its tube a rotted hulk. Today the telescope lives again. During our observing session at Birr Castle, as my eyes grew more adapted to the darkness, I could make out a faint glow just south of the telescope's west wall. I walked over to see a weather-protected computer screen! Lord Rosse would have been astonished to find that his telescope could now be controlled at the push of a button — the Leviathan is now equipped with a hydraulic system, which makes moving the scope a lot easier than it was in his time. Unfortunately, a mechanical glitch in the system had rendered the telescope temporarily inoperable.

Seeing that computer screen really got me to thinking. With a PC already able to provide rough controls of the telescope's aiming and movement, how much more would be involved to add a CCD video camera to it? It turns out that others have had this idea as well. This telescope seems poised to be not just a historical relic brought back to life, but possibly an important tool in public science education and research. Imagine the pride Lord Rosse would feel if he knew his telescope, nearly 160 years later, would be busy searching the sky for asteroids and comets and delighting the world with CCD images! Thanks to the team from Ireland and the UK who helped restore it and to the current seventh Earl and Countess of Rosse, this would be a miracle indeed. JANUARY 2004

# Observing Earth

For those of us who love the stars, weather is always the elephant in the living room we all fear but don't want to talk about. Failure to predict sky conditions accurately is, for most of us, merely an inconvenience. For others, it's much worse. This same failing means also that, three days out, we can't really predict with certainty where a hurricane or tornado will strike.

In recent years much of the improvement in weather forecasting has been largely due to better modeling of the atmosphere with supercomputers. The Clear Sky Clock (www.cleardarksky.com/csk) has been a wonderful tool for amateur astronomers. It uses data collected, modeled, and analyzed by Environment Canada's supercomputers, which are among the best in the world. But like those belonging to the US National Oceanic and Atmospheric Administration (www.noaa.gov), they're only as good as the data that are fed into them. In those data lies the heart of the problem, and in more and better data lies the solution.

From its vantage point in space, the Hubble Space Telescope has given us an unparalleled look at the cosmos. Doesn't Earth deserve the same? It certainly does, according to Conrad C. Lautenbacher Jr., a retired Navy vice admiral who is the undersecretary of commerce for oceans and atmosphere and the administrator of NOAA. As enthusiastic about the interaction of Earth's atmosphere and oceans as we are about the night sky, Lautenbacher brings to his job the energy and vision of a man who understands the urgent need for a coordinated global monitoring network of satellites, ships, aircraft, buoys, and ground-based sensors that's far better than what we currently have.

"We need to fill the many blind spots in our data gathering," he says, "Particularly at ocean depths where right now we get absolutely no data at all. We would have a better ability to predict the courses of dangerous hurricanes, understand more fully the conditions that cause long-term effects like droughts or short-term events like floods, and thus improve our ability to predict them." And for us skywatchers, we'd be able to predict when we'll get a clear night much farther ahead of time than before.

Improved short-term weather forecasts will have far greater impact than just the convenience of scheduling a public star party. Each year in the United States alone, scores of people die and billions of dollars of property, business, agriculture, and infrastructure are lost or damaged as a result of severe weather disturbances. If a global observing system were in place, better forecasts would dramatically minimize those losses.

Although NOAA's present monitoring arsenal is considerable, it still needs improvement. That's why the United States is spearheading

the Global Earth Observation System of Systems, or GEOSS (http://earthobservations.org), a multinational program that aims to develop a comprehensive, coordinated, and sustained monitoring of our planet's weather, climate, and environment.

With a full global observing program in the future, NOAA would offer accurate forecasts for every city and town in the country as much as a week in advance. What's more, it might bring us back, at last, to the biblical story of Joseph, whose dream predicted seven abundant years marked by ample rain and bountiful harvests, followed by seven lean years marked by drought and famine. Once the global observing system is in place and once we fully understand the data it sends us, possibly by 2030 we might make Joseph's dream come true, with accurate decadal forecasts for every city and county in the world.

Wow! Does that mean we'll know if the Texas Star Party will be clouded out a year from now? What about the next total solar eclipse? As we get more observing stations, we'll draw closer to that elusive goal. And indeed it will be as though we've put a Hubble Space Telescope to work on observing Earth round the clock.               SEPTEMBER 2004

# A Voyage to Remember

Flowers are stars of the Earth —
Stars are flowers of the Universe.
— Janet Akyüz Mattei (1943–2004)

At the end of 2004, Gamze Menali of the American Association of Variable Star Observers (www.aavso.org) invited me to speak at the Amateur Astronomy Symposium to be held in Istanbul, Turkey, the following year. The gathering, she explained, was to honor the memory of Janet A. Mattei, who was the AAVSO's director for 30 years. Janet had been a mentor and a friend to countless observers, including me, and we're still in a state of disbelief over her death from leukemia in March 2004.

Janet's impact on variable-star astronomy in these past three decades is without parallel. She took a thriving amateur organization and turned it into a valuable 21st-century research tool for everyone. I don't know of any organization that has managed to accomplish so much so effectively as did the AAVSO under Janet.

Paying tribute to Janet and her astronomical roots in her native Turkey was something we wouldn't want to miss, so late last June my wife, Wendee, and I made the long flight from Tucson to Istanbul via London. Held at Istanbul Kültür University, the symposium was organized by astrophysicist Dursun Koçer. It represented the coming together of the country's amateur and professional astronomers, educators, high-school students, government officials, private university administrators, and the general public. Also present were Janet's family and friends, including her sister Kadem Şenkal.

"Having Kadem felt as if Janet were there with us throughout the meeting," says Menali. Nearly 300 people attended the gathering, which featured some 90 oral and poster presentations and attracted coverage not only in the local newspapers and TV but in parts of Europe as well.

Janet always felt that the life cycle of some of the largest bodies in the universe — the stars — could be mimicked by flowers, among the smallest of Earth's natural wonders. On several occasions Janet's love of flowers would permeate her lectures at AAVSO meetings, and I know that she had hoped to produce a program that would connect her dual loves of flowers and stars. Wendee's *PowerPoint* presentation at the symposium, entitled "From Flowers to Photons," was an effort to perpetuate Janet's passions: seven white flowers gradually morphed into a portrait of the Pleiades; the rare, blossom-laden stalk of an agave, or century plant, was compared

My wife, Wendee (middle), Janet Mattei's sister Kadem Şenkal (far right), the AAVSO's Gamze Menali (standing), and me enjoying a luncheon in Istanbul hosted by Janet's family and friends from her hometown of Izmir.

side by side to a billowing spire of gas and dust in M16, the Eagle Nebula; and a pincushion cactus with a single bloom at one edge mirrored the appearance of T Scorpii, a nova that flared within the globular cluster M80 in 1860 and whose remnants were recently imaged by the Hubble Space Telescope.

For her part, Gamze gave a talk (in Turkish) on the AAVSO's role in bridging the gap between professionals and amateurs and encouraged her audience to participate in monitoring variable stars. Her efforts have already started to bear fruit: one of the attendees, Murat Gezer, volunteered to translate the AAVSO's visual-observing manual into Turkish.

Other speakers discussed topics of concern to astronomers worldwide, such as astronomy education and light pollution. The meeting ended with Haldun, Gamze's husband and fellow AAVSO member, and me engaged in a serious town-hall-style brainstorming session with attendees on the future of the hobby in the country and how to improve amateurs' productivity. We also discussed the possibility of forming a national astronomical organization. During these discussions I realized that, despite the great distance between the US and Turkey, their concerns are our concerns and, despite the language difference, with a little help from Gamze, I left the meeting with a good understanding of the passion these amateurs have for astronomy and for making certain that succeeding generations get an equal chance to enjoy and learn the night sky.

As the Sun began to set on the last day of the conference, participants boarded a boat for a cruise down the Bosporus. This narrow strait separating European and Asian Turkey and joining the Black Sea with the Sea of Marmara has been an important trade route since ancient times. During the cruise, people were dancing, dining, and just having great fun. "It was a voyage to remember," says Gamze, "a voyage between two continents, a voyage between the past and the future. It's amazing how one's heart can experience two equally powerful emotions at the same time: the great joy in sharing the moment with your loved ones and the great sadness in not being able to share it with the departed."

At the end of her presentation, Wendee showed some of her pictures of flowers to the music of an Alabama song that claimed, "I believe there are angels among us." In the gathering dusk that night, as we cruised along the Bosporus, I could imagine Janet in her celestial garden, smiling down on our joyous gathering. DECEMBER 2005

things

# Two on a Tower

In Thomas Hardy's novel *Two on a Tower*, published in 1882, the astronomer Swithin St. Cleeve proclaims, "Whatever the stars were made for, they were not made to please our eyes. It is just the same in everything; nothing is made for man." His new Lady Viviette Constantine, answers, "Is that the notion that makes you feel so sad for your age? I think astronomy is bad for you. It makes you feel human insignificance too plainly."

A profound novelist of English literature, Hardy excelled at pointing out the incongruity between astronomical observing and everyday life. *Two on a Tower* is a story in which earthly happiness competes with the rigor of an astronomical pastime — a choice we amateurs often have to make. Our work goes against the grain; the end of an earthly day is the beginning of ours. As an evening's sky beckons, we forego the warmer pleasures of the night to freeze under the stars. It is a choice that Thomas Hardy portrays dramatically.

"You said you would show me the heavens if I could come on a starlight night. I have come," Vivette announces as she interrupts Swithin's observations. The refractor stands idle and a late 19th-century romance begins. When Vivette shows an interest in astronomy, Swithin cautions her: "If … you are restless and anxious about the future, study astronomy at once. Your troubles will be reduced amazingly. But your study will reduce them … by reducing the importance of everything."

Hardy's astronomical background is solid, complete to the intricate details of constructing and observatory dome atop an ancient stone tower. Vivette purchases a fine new refractor for Swithin's use, and he soon makes a discovery. But when he reports it, he finds himself upstaged by an American fictional equivalent of E. C. Pickering, who was famous for his variable star work. Devastated, Swithin attempts to commit suicide by lying in a ditch during several days of rain. Rescued but near death, his spirit and health do not improve until he hears about a new and brightening comet (patterned after Tebbutt's Comet of 1881).

The romance between Swithin and Vivette deepens, which threatens to steal Swithin from his observing. Vivette fears that marriage would destroy his career, so they wed in secret and agreed to limit their contacts until Swithin achieves success. Although the scheme works at first, a series of small events interferes and threatens to expose them.

Vivette insists that Swithin leave immediately for a tour that would include visiting the refractor at Harvard, seeing the 1882 transit of Venus, and studying the southern sky from the Cape of Good Hope. Astronomy has won out, but at the expense of Vivette's happiness. When Swithin returns, he finds his sky intact but his world in ruins.

Hopefully we amateurs will have more success in integrating our love for the stars and our daily lives. But it is not easy. Because of the long hours and commitment that our avocation demands, many of us might face the same dilemma that gripped Swithin and Vivette.

There *are* ways to enjoy the best of both worlds. For example, why not emphasize family and friends during the week and get that observing fix during the weekend, or vise versa? I do most of my observing in the early

The Southern Cross (upper left) and the beautiful southern Milky Way are invisible to most observers in the Northern Hemisphere.

morning, so I *really* cut against the grain. But working at night while the world passes me by day invites loneliness; it's like always driving the wrong way down a one-way street. Is this loneliness worth it? Each new comet, each outburst of distant cataclysmic variable, each stunning aurora assures me that it is.

There are many reasons why we spend time under the stars. By looking skyward, we see the stability that exists in something so immeasurably greater than ourselves. If we are serious enough, we can live in our observatory towers, sometimes physically but always in spirit. Only we can relate to canceling a social evening because a cold front has just revealed the first clear sky in weeks. When our non-observing friends ask why, we have to explain that the vagaries of the weather and the darks of the moon are our real obligations.

"There is a size at which dignity begins," Swithin declares. "Further on there is a size at which grandeur begins; furtheron a size at which awfulness begins; further on a size at which ghastliness begins." This astronomer's world is so demanding, that he can't help but to discourage Vivette from studying it. "If you are cheerful, and wish to remain so, leave the study of astronomy alone. Of all the sciences, it alone deserves the character of the terrible."                                        JULY 1990

# Tombaugh's Star

On the morning of May 25, 1932, Clyde Tombaugh was blinking two large photographic plates he had exposed the previous year. They were of a region in Corvus. More than two years had passed since his momentous discovery of Pluto. Although he hoped to find more planets, his all-sky survey at Lowell Observatory was yielding "only" asteroids, variable stars, and other curious objects.

At 11 o'clock Tombaugh suddenly stopped. There was a bright 12th-magnitude star on one plate where none appeared on the other. A plate taken with a different camera the same night (March 23, 1931) showed it too. Tombaugh reported the discovery of a probable nova to fellow as-tronomer Carl Lampland, but somehow word of it never escaped the ob-servatory. The star's images remained buried in the emulsions for almost 60 years; even Tombaugh forgot about them.

While preparing a biography of Tombaugh in 1988, I visited the Lowell archives to inspect the detailed notes he had written on the back of each plate envelope. There I learned of this discovery made and then lost.

The star's image looked real enough on the original plate. "But wouldn't it be more convincing," I thought, "if I could find records of the same outburst taken with telescopes elsewhere." So during the sum-mer of 1989 I searched the gigantic plate archives at Harvard Observa-tory, fingers crossed that their cameras had recorded the event. Alas, the nearest plates had been taken one week before and five weeks after the outburst, and both showed just empty sky.

Not ready to give up, I considered that this star might be a *recurring* nova and that the Harvard collection might contain an image of another outburst. I began a systematic search of the region, expecting to scan countless plates with no success. Fifteen plates later the search paid off — the star was there again! But that was too easy! How often did this star erupt? With mounting interest I continued my search through the plates Harvard had of the Corvus region. Three days passed before I finished, but in that time I found nine additional outbursts.

Thinking it was time to relax and watch the accolades roll in, I report-ed my findings to Brian Marsden of the Central Bureau for Astronomi-cal Telegrams. Marsden agreed that this star was very likely a new cata-clysmic variable, a fainter version of famed SS Cygni or U Geminorum. These eruptive stars are binary systems in which a hot white dwarf draws hydrogen off its larger, cooler neighbor. Eventually the accumulating gas ignites in a brief thermonuclear explosion, then the accretion process re-peats itself. Marsden suggested that very likely such an explosion would occur again. Why not try to observe the next event visually?

In late November, as the predawn curtain rose on Corvus, I made the first of about 60 visual checks with my 16-inch reflector from my back-yard in Tucson. Since the star is 19th magnitude at minimum, (much too faint for my telescope), each night's check revealed only blank sky where I hoped to see the star. The project challenged me, for a single cloudy night could cause me to miss the outburst completely; some nights I would wait, telescope and eyepiece ready, for a tiny, well-place hole in the clouds.

On the morning of March 23, 1990, 59 years to the night since the outburst Tombaugh detected, I looked through the eyepiece and saw the star at magnitude 13.5. After such a long wait, the thrill was akin to discovering a comet. I alerted a network of observers primed for this occasion. Gary Rosenbaum on Kitt Peak arranged for a spectrum to be taken within an hour of my phone call. Brian Skiff at Lowell contacted Rob McNaught, who arranged for a spectrum from Australia. Marsden reported the results on International Astronomical Union *Circulars* 4983 and 4987.

The work on this star is just beginning. Since it has a decent "outburst histoy" it is a good target for close monitoring by amateur variable star observers. But be sharp. One night after maximum, the star faded to magnitude 14.5. If this behaviour is typical, an entire outburst can be missed thanks to two cloudy nights. Tombaugh's new variable star in Corvus is now included in the program of the American Association of Variable Star Observers.

Why observe this star? For one thing, its range in brightness is about 7 magnitudes, unusually large for a cataclysmic variable. Also, while analyzing the spectra of the March, 1990, outburst, Lowell astronomer Steve Howell noticed the presence of hydrogen-alpha emissions (*Circular* 4987). These are usually absent when cataclysmic variables erupt. More observations are needed when the star is in the early stages of an outburst, so that more spectra can be obtained. Last but certainly not least, we owe thanks and credit to Tombaugh, the former Kansas farmer whose long years in Flagstaff produced a data set that still yields interesting objects to challenge and delight us all.                    JANUARY 1991

### Author Update:

*If you're interested in observing Tombaugh's star (officially known as TV Corvi), contact the AAVSO (www.AAVSO.org). But to monitor it you'll need a large telescope, because even at it's brightest, the star is barely 13th-magnitude. In February 2005, I had the rare privilege of imaging TV Corvi just as it began its three-hour climb to maximum brightness. This star, which usually brightens no more than once per season, continues to give me many satisfying nights.*

# Tombaugh's Telescopes

In 1926 a young man, eager to explore the starlit skies over his Kansas farm, tried making his own telescope. The 8-inch glass he ground was less than perfect. Mounted in a 7-foot-long wooden box, the mirror barely formed an image and nearly cost young Clyde Tombaugh his interest in astronomy.

Yet Tombaugh spent the next year tenaciously trying to improve his mirror-making techniques. His first worthy accomplishment was a 7-inch reflector, followed a year later by a 9-inch, through which he studied the planets. The meticulous drawings he made at the telescope helped him to get a position at Lowell Observatory, where he started his search for trans-Neptunian planets. In 1930 he made history with the discovery of Pluto, our solar system's ninth planet.

In his spare time Tombaugh continued to design telescopes, including a 5-inch f/4 reflector, the first rich-field telescope built in the United States. His skills proved valuable when he left Lowell Observatory in 1946 for the White Sands Proving Grounds in New Mexico. There the US Army was testing V2 missiles captured at the end of World War II, and Tombaugh was to design and operate telescopes that could track them after launch.

Before Tombaugh arrived at White Sands the favored launch time of a V2 was near 11 a.m. Tombaugh, however, quickly pointed out that, as any solar observer knows, the midday sky is highly turbulent, making

Clyde Tombaugh (in 1995) with his 9-inch reflector in Las Cruces, New Mexico.

photography of objects in the sky difficult. Indeed, the early images of V2's obtained at White Sands were a joke, the rockets appearing as unrecognizable fuzzy streaks. The officials questioned Tombaugh's attempt to explain this "seeing" effect. "Turbulence? What's that?" But several more midday launches brought equally poor results.

A recalcitrant V2 finally made Tombaugh's day. As delays pushed the launch later into the afternoon Tombaugh grew more hopeful. Finally the rocket took off into a deep blue sky, with the Sun low and behind the camera. The pictures were fantastic, clearly showing the V2 110 miles away. From that day forward Tombaugh's photographs became an increasingly valuable record. Even the paint pattern on each rocket was changed from black and yellow-green to flat black and flat white, with diagonal stripes for higher contrast, providing clearer images in the telescopes.

Tombaugh triumphed again when his photographs solved a riddle that had long plagued the German-designed V2's: occasionally the rocket would gyrate as its fuel ran out. The now-distinct photographs showed that the last drops of fuel were firing off to one side. A bit of primacord ignited a fraction of a second before fuel exhaustion, cleanly cut off the flow to the engines.

By 1950 the cold war had entered a dangerous phase. With much of the country fearing imminent Soviet attack, defense strategists envisioned a series of Nike antiaircraft missiles ringing the major cities of the United States. White Sands was selected to become the principal Nike test facility, with Tombaugh responsible for creating a telescope to track them. His Intercept Ground Optical Recorder (IGOR) was a 16-inch f/6 Newtonian telescope designed to capture the details of a missile intercept a minimum of 5 miles away.

The experiment began as an unpiloted B-17 bomber was sent aloft. An operator on the ground controlled the plane's course by radio as it climbed into the sky. Joe Marlin, a telescopist trained by Tombaugh, operated the IGOR and its camera. With a roar the Nike then left its launch pad, approached the plane, and scored a direct hit on one of the engines. The plane's starboard wing collapsed, and the plane fell apart.

In 1952, with attention turning to the possibility of space travel, Tombaugh proposed that the Army conduct a search for natural Earth satellites 13th magnitude or brighter. After all, what if the planet were surrounded by lots of tiny satellites that could pose a danger to spacecraft? Using a Schmidt camera at Lowell Observatory and two fast lenses in Quito, Ecuador, Tombaugh and his research associates scanned the sky for satellites at various distances from the Earth. Their tracking varied from a sidereal rate to an extreme 400° per hour! Nothing was found. The program's timing was marvelous, ending just as Sputnik was launched late in 1957. With the blanket of artificial junk now in the sky, there is no longer any hope of detecting a tiny natural satellite.

For the decade following World War II, Tombaugh detoured from his love of astronomy. But thanks to his equal passion for telescopes, he became one of the few astronomers who participated in the dawn of the space age. APRIL 1992

### Author Update:

*Clyde Tombaugh passed away on January 17, 1996, at the age of 90.*

# My First Telescope

There's a certain magic associated with one's first telescope. Mine was a 3½-inch f/10 reflector with a sleek black tube. My Uncle Sydney bought it from the Hayden Planetarium gift shop while visiting New York in the mid-1950s. He originally intended it for his children, to get them excited about astronomy, but they weren't interested. The scope remained unused until one memorable afternoon in 1960 when I walked into our living room to see Uncle Sydney explaining its use to my parents.

The night that followed was golden. Together in the backyard, my parents and I set up the telescope, which we named Echo, after the newly launched communications satellite. Two bright stars shone in the southern sky, and I turned the telescope to the more conspicuous one. After aligning my eye with the small eye lens, which took a few minutes, I finally saw a large, round blob of light with a hole in the middle. With a push of a tube here, and a turn of a knob there, I quickly learned about focusing. The doughnut-shaped blob gradually shrank to a beautiful disk with two belts across its middle, accompanied by four small stars. Thus, Jupiter became the subject of my magical first view through a telescope. My parents and I had neither star charts nor planetary tables. Jupiter just happened to be there, and I just happened to focus on it.

Almost exactly 34 years later I again turned Echo, with its same 60-power eyepiece, toward Jupiter. Now, as then, two large belts crossed the planet's equator; but this time there was also a series of dark spots high in the planet's southern hemisphere. They are the remains of a comet gone fatally awry. Echo, Jupiter, and I had come full circle.

Nowadays most small scopes found in department stores (where many uninitiated buyers first see them) are of inferior quality compared to those of my generation. Needless to say, the majority of experienced amateur astronomers today discourage anyone from buying a small-aperture telescope from such outlets. Still, this does not help the many children who already own one of these instruments. It's their lasting interest in the sky that's in jeopardy.

After tinkering with various telescopes of questionable character, I discovered a way to make even the cheapest ones more effective. It turns out that most of these beginner scopes have a crazily tiny eyepiece, one that offers little eye relief and leads to much frustration. Getting one's

eye lined up with that lens is like trying to thread a needle in the dark. A better eyepiece, one with a wider field, eliminates this problem!

I tested my theory by

Not only do I still have my first telescope, I continue to use it. Echo makes occasional appearances at star parties, and I frequently employ it as a solar telescope.

replacing Echo's 0.965-inch eyepiece holder with a standard 1¹/₄-inch low-profile one. This size was chosen for most American-made eyepieces because, in the heyday of amateur telescope making, that was a standard tube diameter. As I carried the telescope outside, I grabbed the first eyepiece I saw — a 25-mm Kellner, inserted it, and looked. Without doubt, the view was by far the best Echo had ever delivered in more than three decades.

Intrigued, I then switched to some larger and more expensive eyepieces that offered much wider fields of view. Although Echo almost collapsed under their weight, the image got better with each ocular as long as I stayed with bright objects like the Moon or Jupiter. Then it dawned on me that for a bright object the quality of a scope's primary mirror or lens is not especially critical, provided it has a long focal ratio — say, f/10. Such a mirror doesn't even have to be parabolized; any long-focus objective will work well. The cheap beginner's telescope lives!

Besides their generally better optical quality, larger eyepieces have the added advantage of wider eye lenses. I found that children have an easier time looking through a wide eye lens, especially in the dark. If you're planning a star party for children, consider holding it when the Moon is in the sky, so you can see if the bright focused beam of moonlight is entering the child's pupil.

Another common problem with small telescopes is their wobbly mountings. You can improve them by making sure all the screws are snug and then spreading the tripod legs out until the chain holding them is taut. You can also hang some weights, like a gallon jug of water or a sandbag, under the base of the scope's mount to make it rock-steady. These telescopes won't be the pride of a star party, but they should give their young owners less jiggly views of the Moon and bright planets.

I take Echo along every time astronomer Larry Lebofsky and I put on an ARTIST program at a Tucson school. They seem to enjoy knowing that the telescope I started with when I was about their age can still break the bonds of Earth and fly me to new worlds beyond. NOVEMBER 1995

# A Sonnet for Columbia's Seven

*The ship stands tall, the sky looks down so soft.*
*A second Sun lights up the launching site*
*And slow, and straight, Columbia roars aloft*
*Past birds, past planes, past air, to space and night.*

*Around the Earth, six million miles she sailed*
*With seven lives, yes seven dreams and more.*
*They saw a world whose nations' borders paled;*
*They saw a world unstained by crime or war.*

*Blue world, alone in a sea of space and time*
*For two weeks joined by a tiny, special moon*
*Looked out as one to far-off worlds sublime —*
*But time had come to head for home — so soon.*

*So far they flew, yet so much more is nigh;*
*So short their time, too soon for souls to fly.*

The crew of *Columbia's* STS 107 mission making its way to the Kennedy Space Center's Launch Pad 39A Shuttle's January 16th liftoff

**As I turned my telescope to the stars,** seven astronauts turned their Space Shuttle toward Earth. It was Saturday morning, February 1st. For me, it was the end of a clear and wonderful night; for them, it was the end of a flawless two weeks in space. As dawn broke, I closed the observatory roof and went inside the house. Thirty minutes later *Columbia*, which had been flying without incident since April 1981, passed over well to our north. Observing from Flagstaff, Arizona, astronomer Brent Archinal and his colleagues caught sight of the shuttle's dramatic reentry.

"We saw a bright point of light of about magnitude –1 appear roughly 10° high in the northwest, and as it flew over it climbed to about 30° and brightened to –2," Archinal recalls. "With the naked eye it was a brilliant starlike object with about a 30°-long whitish plasma trail. Through binoculars the object appeared as a tiny fireball whose appearance was changing rapidly. At no time did I see more than one object."

Within the next five minutes, however, that story would change. While plunging through the atmosphere at 18 times the speed of sound (20,000 kilometers per hour) at an altitude of 62,700 meters (207,000 feet) over east-central Texas, *Columbia* suddenly broke apart, resulting in the loss of both vehicle and crew. I saw a pretty predawn celestial scene, Archinal saw a stunning fireball, and observers farther east saw a catastrophe.

In the aftermath of the *Columbia* tragedy the inevitable question has returned: Do humans belong in space? Even among my own friends, opinion is divided; maybe now is the time to put all our resources into the robotic exploration of the solar system, some say. But I don't think that this is practical. If human exploration of space were to stop, I believe that funding would very quickly dry up for all forms of space exploration. But there's more. Space exploration is part of our lives; it certainly has been a part of mine since I stayed home from school to watch Alan Shepard's suborbital flight on May 5, 1961. As his small Redstone rocket lifted off its launch pad, it took my dreams into space with it, and they've stayed there ever since.

"To honor the legacy of the *Columbia* astronauts we have made a solemn commitment to their families to find the cause of the shuttle accident, correct what problems we find, and safely move forward with our work," said NASA administrator Sean O'Keefe during a memorial service at the National Cathedral a week after the tragedy. "Motivated by our mission goals of understanding and protecting our home planet, exploring the universe and searching for life, and inspiring the next generation of explorers, we will make good on this commitment."

For most of us NASA is a close and dear friend; it's that personal. I remember space successes joined with great observing sessions. During my first all-night session on August 12, 1962, I saw 112 Perseid meteors and listened to news reports of the first rendezvous in orbit between Russian cosmonauts Andrian Nikolayev and Pavel Popovich aboard Vostok 3 and 4, respectively, as they passed within 5 km of each other. Just seven years later, in a crowded auditorium at Camp Minnowbrook in New York's Adirondack Mountains, I watched as Apollo 11's Neil Armstrong and Edwin "Buzz" Aldrin took their first steps on the Moon. In 1990 I watched with friends as *Columbia*'s Astro 1 mission returned to Edwards Air Force Base in California's Mojave Desert after observing many targets in the ultraviolet, including a comet I had found.

Probably my most intense NASA experience was in 1994 and 1995, when a fleet of spacecraft, which included the Hubble Space Telescope, Galileo, the International Ultraviolet Explorer, the Extreme Ultraviolet Explorer, and the Astro 2 telescope array (aboard the Space Shuttle *Endeavour*), observed the effects of a comet collision with Jupiter.

But if we share in NASA's joys, we must also share in its sadness. On Friday evening, January 27, 1967, I was at an astronomy club meeting when we learned that Apollo 1's three astronauts had perished in a flash fire during a prelaunch test at Florida's Kennedy Space Center. Exactly 19 years later, on January 27, 1986, I took a CCD image of Halley's Comet less than 30 seconds before it was to set at dusk. A few hours later I awoke to those terrible images of *Challenger* exploding in the sky.

Even though I don't feel a need to go into space myself, I remain an unabashed proponent of the space program. My predawn observing sessions, with just me, the sky, and my telescope, get me as close to space as I feel I need to be. As I write this, however, human exploration of space goes on with three people aboard the International Space Station. While the ISS is a noble effort, it should not be the end of space exploration. Let's get back aboard the shuttle as soon as possible, but let's not stay there. We should design a spacecraft that can get us farther into space. After all, those tall Saturn V boosters lying on their backs at Cape Canaveral and other NASA sites are not about to jump up and take us to the Moon again. Those glory days are over, but could they be replaced with new glory days? We can go to the Moon, for keeps this time, and observe the sky from a base there. I hope my grandchildren will witness a successful human voyage to Mars, the planet that has beckoned us since the time of Percival Lowell.

In the meantime let's keep up our robotic missions. The New Horizons mission is a wonderful probe that deserves to go to Pluto before the planet's atmosphere freezes up for more than 200 years. We need to go on: exploring new worlds, whether by humans directly or through their robotic eyes and ears, is what the space program and our future are all about. It's also a vicarious way of getting to those distant shores ourselves.

"I wanted to ride on that arrow of fire up to the heavens," the late singer John Denver once wrote. "They were flying for everyone. They were flying for me."                                     JUNE 2003

### Author Update:

*On July 26, 2006, the space shuttle* Discovery *was launched from NASA's Kennedy Space Center, ending a two-and-a-half year post-*Columbia *hiatus.*

# Adventures with
# Mr. Schmidt's Telescopes

How many times have we heard from beginners who give up after unpacking their telescope, setting it up, and not seeing anything through it? Chances are these people either returned their scope to the store or stored it in a closet, never to use it again.

Leslie C. Peltier (1900–80), one of the 20th century's greatest amateur observers, wrote about his first look through his 2-inch telescope in his autobiography, *Starlight Nights:*

> I then removed the brass cap that covered the lens, pointed the telescope toward Grandpa's house across the field and, with the feeling that this was one of life's great moments, I looked in the eyepiece. I saw only a blur!
>
> In that first frenzied hour I learned many things about telescopes. First of all I learned that they need to be focused. That an adjustment of the distance between the large lens and the eyepiece — made by sliding in or out the smallest drawtube carrying the eyepiece — is necessary in order to make the image sharp and clear when seen in the eyepiece. I learned that this adjustment varies with the distance of the object that is viewed, that the more remote the object the nearer must be the eyepiece to the lens. And to my great delight, when this adjustment had been made, I learned that Grandpa had flies that sunned themselves on the siding of his house and dark green moss that grew between the red bricks of his chimney.

Peltier could simply have packed up his telescope and returned it. Then he would never have gone on to discover a dozen comets and two novae nor make more than 132,000 variable-star observations.

Any new telescope, especially a first one, needs to be broken in with care. The key is to work with the telescope. Find out how it performs. Let it teach you, the observer, how it can show you the heavens.

Even an experienced observer and telescope user like me can sometimes run into trouble operating an instrument. Take, for example, the story of Ophelia 1 and 2, my two Celestron 8-inch (0.2-meter) Schmidt telescopes, which I had acquired in 1988. Their focus had been fine-tuned by Epoch Instruments using laser collimation. The company did a good job, and both scopes worked very well.

The Schmidt telescope, incidentally, was invented in 1929 by Bernhard Schmidt, a brilliant Estonian optician whose life was the cover story for the November 1955 *Sky & Telescope*. He was able to grind and polish lenses and mirrors manually to a high precision despite having only one arm (he lost his right forearm in his youth while experimenting with explosives). His invention is actually a photographic camera specially designed to capture large swaths of sky with razor-sharp star images across the entire field. Its optics feature a thin corrector plate and a spherical primary mirror with a short focal ratio, typically f/2 to f/3. The film is pressed flat against the film holder, which has a curved focal plane to match the curvature of the primary mirror.

After Gene and Carolyn Shoemaker and I ended our asteroid and comet survey at Palomar Observatory in 1996, we set up Ophelia 1 and 2 in Flagstaff, Arizona, as our Shoemaker-Levy double cometograph. But transporting the telescopes up to Flagstaff and replacing the original 35-millimeter film holder with a 55-mm one that uses a thicker emulsion necessitated refocusing the scopes. With Celestron no longer manufacturing Schmidt telescopes and Epoch Instruments not servicing them anymore, my close friend Robert Goff, who was a master optician, came to the rescue. In his Tucson optical laboratory Bob tweaked the focus of both cameras using another telescope focused at infinity and pointed directly into each Schmidt. For the first few years of our comet-survey program we used the Schmidts both in Flagstaff and at our home near Tucson. While Ophelia 1 retained her focus beautifully, Ophelia 2's images got soft and, over time, her star images turned into little doughnuts. Bob decided to refocus the scope more critically by taking it under the stars.

In Celestron's Schmidt telescope, the film holder is held in place by a

Here I am (kneeling) with my black Celestron 8-inch (0.2-meter) f/1.5 Schmidt telescope named Ophelia 2 at my backyard observatory in Vail, Arizona. We're joined (from left to right) by my wife, Wendee; Dean Koenig; and Robert Goff's widow, Valerie.

spider assembly, which in turn is supported by three Invar-alloy rods running down the length of the tube. Because of its physical configuration, you can't look through a Schmidt telescope to verify its focus visually. Instead you have to check and recheck the image quality photographically. To adjust the telescope's focus, Bob had to loosen the nuts both on the mirror side and on the corrector-plate side of each rod — six nuts in all — tweak the focus, and retighten the nuts before taking a short exposure. I developed the film quickly, and we examined the still-wet negative with a magnifier. After making an adjustment we took another exposure. We repeated this procedure dozens of times, each session taking more than an hour. Since the system's photographic speed is a lightning-fast f/1.5, a movement of $\frac{1}{1000}$ inch makes the difference between doughnuts and pinpoint star images. The first 15 or so exposures were to collimate (align) the corrector plate before we could even begin the task of focusing!

We spent several evenings, spread out over months, on the focusing process. Each time we would get closer, then farther, from focus. Focusing evenings became impromptu star parties for Bob and his wife, Valerie, and me and my wife, Wendee. By the end of 1999 we actually started to wonder if Ophelia 2 would be operational by the start of the next millennium.

With Bob's health failing, we stopped using Ophelia 2 for our patrol work. Two days before Christmas 2001, Bob's heart finally gave out. Shortly after his memorial service, my friend Dean Koenig (see page 73) was reminiscing about Bob's considerable accomplishments in the optical field. We realized that we needed to finish the job of focusing Ophelia 2. It was now a quest — it's what Bob would have wanted us to do.

Dean and I continued the monotonous task of focusing Ophelia 2. Each night we took perhaps nearly a dozen pictures. The agony was that on several occasions we *almost* achieved focus — just one edge of the field would be slightly off, and, in our attempt to correct it, the opposite side would become blurry, and we'd get farther and farther away from focus. "We were perfectionists," Dean admitted. One night the sky began to get very hazy, as if a strong wind were blowing dust into the air. We were baffled because there was no wind! The next morning we found out that a huge dust storm from Mongolia's Gobi Desert had blown dust across the Pacific, drifting over Arizona and hindering our efforts.

Since I was developing the film right out in the observatory, I removed my Gibeon meteorite wedding ring to prevent contamination from the film's developer. The observatory ate the ring one night, and I haven't seen it since.

Finally — on a heroic night in the spring of 2002 — Dean and I got Ophelia 2 to a pretty good focus. After three long years we finally put the scope back online, and Carolyn, Wendee, and I have been taking pictures with it ever since. I might have lost my ring, but the time I had spent with Bob and Dean was precious. I wouldn't give it up for anything.

So when you take your brand-new telescope outdoors for the first time, spend a few moments getting acquainted with it. It's worth it. Just as we're now enjoying nice images with Ophelia 2, may you enjoy years of discovery with your new — and personally focused — telescope.     NOVEMBER 2003

# Further Adventures
# with Mr. Schmidt's Telescope

In the previous column, I described the long, painful process that was involved in focusing the 0.2-meter (8-inch) f/1.5 Celestron Schmidt telescope used in my photographic hunt for comets. When my friends and I finally achieved focus after many, many nights of careful experimentation, we thought we were done. But that was not to be. During the six years of our efforts, from 1996 to 2002, digital technology evolved very rapidly, and eventually the time came for my wife, Wendee, and I to make the switch from film to CCD imaging. So in June 2003 we loaded the last sheet of hypersensitized Kodak Tech Pan 4415 film into the Celestron Schmidt and took an 8-minute exposure. That shot turned out to be a lucky one — we captured the Hubble Space Telescope as it passed over our backyard observatory in Vail, Arizona.

The events leading up to our switch to digital imaging actually started in 1999, when Scott Roberts of Meade Instruments called to say that Meade was offering me a grant of its beautiful 12-inch LX200SC Schmidt telescope. It would be the largest such instrument I've used since Palomar Observatory's venerable 18-inch Schmidt, which was the workhorse in the asteroid and comet survey that I shared with Gene and Carolyn Shoemaker from 1989 to 1996. News of Meade's 12-inch version didn't come as a complete surprise since Scott and Meade founder John Diebel had told me of their plans for a big Schmidt scope when they visited Palomar just before the Comet Crash of 1994. By the end of 1999, one of the company's units was headed for us.

The 12-inch is so big, however, that it necessitated a major expansion of our backyard observatory. I remembered how, when I was a teenager in Montreal, our local observatory had a piggy bank named Obadiah, and that every few months the club's newsletter would announce how "Obadiah, the Observatory Pig, went shopping." Taking up so much room even in our expanded building, we dubbed the 12-inch "Obadiah."

At f/2.2, Obadiah was a little bit slow for our photographic search — it would require exposures of up to 15 minutes to reach the telescope's magnitude limit. So when the opportunity came to try to convert it to CCD imaging, I accepted the challenge right away. That opportunity arose while Wendee and I were hosting a *Sky & Telescope* tour group at our home. One of the guests was Pierre Schwob, a California entrepreneur and astronomy enthusiast who took an immediate interest in our telescope, and who, months later, awarded us a grant to attempt the conversion.

Classical Schmidts use film that can easily be flexed to match the

My grandson Matthew and I check out Obadiah, the telescope I use to hunt for comets at Jarnac Observatory in Vail, Arizona.

telescopes' curved focal planes so that star images across the emulsion appear sharp. But a CCD camera's silicon chip is not flexible. Richard Buchroeder, whose optical genius I had known for decades, was my next stroke of good fortune. He offered to design a field-flattening lens for Obadiah, and master optician Thom Peck ground the glass perfectly.

But what kind of CCD camera should I use and how should it be mounted? Dean Koenig of Starizona recommended the STL-11000M, which had just been introduced by the Santa Barbara Instrument Group. SBIG's large chip would cover a 3°-by-2° swath of sky in a single exposure, but the camera's large, boxy housing made it too bulky to mount inside the telescope. Fellow comet discover and instrumentation guru Roy Tucker had a brilliant idea: he disassembled the STL-11000M, extracted the CCD chip, and mounted it exactly where the Schmidt's film holder used to be. The rest of the camera electronics remained in the original housing, hanging underneath the telescope tube. Cooling the chip is done with a thermoelectric system and a small pump that circulates antifreeze through plastic tubes. Electronics expert Dail Terry added a double-fan heat exchanger to increase the setup's cooling efficiency.

The whole conversion, from the final film exposure with the Celestron Schmidt to Obadiah's first CCD image, took about two years. But Obadiah was still not ready for prime time. Recalling the multitudinous nights of gentle coaxing of the Celestron Schmidt to attain perfect focus, Dean and I began a series of more than 300 exposures with Obadiah, each lasting 30 seconds, taken over four nights. We would examine the quality of the star images at each corner of the frame on the computer monitor and then gently and cautiously adjust the telescope. The process was painfully slow. On the fifth night, we still had not achieved perfect focus but were close enough to set Obadiah loose on a comet search.

Thanks to a talented team of experts, I can now say that my goal to hunt for comets with a CCD-equipped Schmidt telescope has finally succeeded. Combining the classic 1930s optical design of Bernhard Schmidt with modern electronics, Obadiah now gracefully scans the sky automatically, courtesy of Bob Denny's *ACP Observatory Control Software*. The telescope is also capable of taking hundreds of images each night down to 18th magnitude.

All that's missing now is a comet.                    JULY 2006

objects & events

# A Sunrise Total Eclipse

"Anybody who tries to see a total eclipse of the Sun in the North Atlantic at sunrise," someone told me, "has to be crazy." At first I didn't realize that the great European eclipse of August 1999 would have anything at all to do with Nova Scotia, the Canadian maritime province that clouded me out the last time I tried to see a total eclipse there. That one happened on March 7, 1970. The Moon's shadow rushed up from the southwest that cloudy day, and I was really impressed with how dark the landscape had become due to the eclipse and clouds. Could it be that the Nova Scotia weather might just work for us this time?

If we were to believe the climate statistics, we had less than a 30 percent chance of clear skies at our viewing site in the North Atlantic on August 11th. And that was not taking into account fog or other local conditions that would prevent us from seeing the totally eclipsed Sun less than 4° above the eastern horizon. But having attended Nova Scotia's Acadia University for four years, I also knew that the area was capable of offering fine weather at this time of year.

So my wife, Wendee, her family, and I took our chances, and on August 8th we joined 670 fellow eclipse chasers aboard Regal Cruises' *Regal Empress* in New York. That evening we met with Captain Peter Schaab, who understood our needs and showed us a detailed map of the eclipse track southeast of Nova Scotia.

By the time we arrived in Halifax on August 10th, the sky was clear and the sea smooth: the Nova Scotia I remembered. But there were a few high cirrus clouds that got some of us worried. "Horsetail cirrus!" Roy Bishop called out. An accomplished amateur astronomer and the editor of the Royal Astronomical Society of Canada's *Observer's Handbook*, Roy also knew the weather patterns of his native province. "These clouds should not present a problem," he assured us.

In order to give us plenty of time to reach the eclipse site, the *Regal*

*Empress* had to depart Halifax by 4 p.m. Right on time, the ship steamed out of Halifax harbor and headed southeast. As Roy, Wendee, and I watched in disbelief, the ship turned south, then southwest, and finally west. The ship had made a wide U turn and was heading slowly back into the harbor! What could possibly have gone wrong? On this beautiful day, were we to miss the eclipse because of engine trouble or something else? As the *Regal Empress* entered Halifax harbor again, we felt our chances of viewing the eclipse drop by the minute.

The problem, it turned out, was far less serious. Two passengers had missed the ship! The *Regal Empress* sailed partway into the harbor, met a small tender, and two people climbed, with some embarrassment, up a rope ladder onto the ship. Turning southeast again, the *Regal Empress* once more began in earnest her voyage into darkness. With nightfall came a clear and magnificent sky. We stayed outside for a while, watching occasional Perseids and a Milky Way so bright that its light reflected onto the sea. Other than an intermittent fleecy cloud, the sky was perfect.

At 4 a.m. on Wednesday, August 11th — the Big Day — I was up on the ship's forward deck just below the bridge. We were definitely not sailing beneath trans-Atlantic air routes, nor were we in heavily traveled shipping

lanes. There was just one other ship far to the north, and as I watched, her lights disappeared over the horizon. There we were, one small vessel amid a vast ocean and a sky so bright, as Grandma used to say, that the stars hung down like baseballs. I thought of poet Samuel Taylor Coleridge's an-

A regal total solar eclipse. On the morning of August 11, 1999, 670 eclipse aficionados rendezvoused with the Moon's shadow aboard the *Regal Empress* in the North Atlantic some 200 miles southeast of Nova Scotia. This photo of the diamond ring marking the start of totality was taken by the ship's photographer.

cient mariner — "Alone, alone, all, all alone;/ Alone on a wide, wide sea." And upon this wide sea the Sun would rise, then vanish in total eclipse.

As dawn approached I set up my telescope, Ophelia; the sea was calm enough that I could hunt for comets quite comfortably. And effectively, it turned out; after 15 minutes I actually discovered a comet! Not a new one, unfortunately — it turned out to be Comet Lee, which recently crossed into the morning sky on its own trek through the ocean of space. Now dawn was advancing, and with it, low clouds on the eastern horizon! I woke up Wendee, and by 5:30 our group was assembled on the ship's starboard side. We were now well within the path of the Moon's shadow, and the ship deftly swung south to take advantage of a large clear area between low clouds to the northeast and southeast.

The sky continued to brighten, but the light level began to drop just before sunrise, and we saw a sight that would have stunned Coleridge's mariner: a thin, alien-looking crescent Sun rising right out of the sea.

Finally positioned, Captain Schaab swung his ship so that her starboard side faced east. The sky was now darkening fast. We could see tiny crescent Suns reflected on the sea. "One minute to totality!" we heard from the loudspeaker. But that timing was off. Besides, so much happens in the last minute before totality that a countdown is not necessary. The sky was darkening with the speed of a celestial dimmer switch. Patsy Tombaugh, widow of Pluto discoverer Clyde Tombaugh, was standing near us and saw a single shadow band race by. Wendee took a picture of the Sun hanging a mere 3.8° above the horizon, and as she prepared to put her eclipse viewer back on, she yelled out, "David! Diamond ring!"

Less than a minute after the Moon's long shadow first touched Earth's surface, it enveloped us. And a sunrise total eclipse, we quickly learned, is unlike any other. Like the minute hand of a huge cosmic clock, the 30-mile-wide umbra visibly swept across the sky during the 52 seconds of totality. And the solar prominences, seen so close to the horizon, took on a very deep color. Wendee likened them to dazzling rubies! The corona, roughly circular as expected at a time of heavy sunspot activity, also looked quite red. We got a sense that the Sun loomed extraordinarily large over the water — a solar counterpart to the famous "Moon illusion."

At midtotality, I decided to check out the prominences and corona with my Ophelia. Never again would my telescope have another observing session quite like this: first stumbling across a comet, then two hours later witnessing a total eclipse of the Sun. But there was no time to think about it, for as I looked up the Moon's shadow was racing east, and fast. A brilliant spark of light, a fading corona, a chorus of screams and applause, and nature's greatest show was over. Alone on a wide, wide sea, a single, long blast from the ship's horn saluted the Sun, the Moon, the ocean, and the unforgettable drama of a total eclipse.                    JANUARY 2000

# Meteor Nights

Meteor observing is the purest way to watch the sky. With no telescope or binoculars involved, it's just you and the changing sky. With passing time you can watch the constellations crawl along the celestial vault as the meteor rates change hour by hour.

Since I first looked skyward one summer evening 44 years ago and wondered what that small, momentary streak of light was, I've always been interested in meteors. And then on August 12, 1962, I watched my first meteor shower from the dock of Grandpa's country home at Jarnac Pond, north of Montreal. "Are you going to sit there all night long?" he asked. "Yes," I replied, "until dawn." I explained to him how, as Perseus rose higher, we would see more meteors; the shower would be strongest just before dawn. That meant, I insisted, that I was going to stay up all night.

Knowing my reputation for being a klutz, Grandpa built a makeshift fence on the dock so that if I fell asleep I wouldn't fall into the lake and drown. As the night wore on the clouds gradually dissipated, and I was absolutely thrilled to see the prediction come true. With each passing hour the number of meteors increased. One of them appeared to split in two; others left beautiful sparkling trails. Alone by the lake that night, I was treated to my own personal show of fireworks in the sky. When dawn finally broke I had logged 112 Perseids.

As the years have gone by, I've never lost my love for meteors. The shower that most occupied my thoughts in recent years was the Leonids. Would they really come in vastly increased numbers as they trailed their parent comet Tempel-Tuttle?

In November 1995 I saw a magnitude −8 Leonid dart through Orion, casting shadows on the ground. A year later I was part of a special meteor watch in Carbondale, Illinois. The large group that had gathered for the event was disappointed by clouds. In 1997 the meteors seemed to turn themselves on and off in bursts of activity. And the following year my wife, Wendee, and I watched more than 200 of them fall over Mauna Kea, Hawaii.

With fellow amateur astronomers Bob and Jean Citkovik and Alan Stern and his daughter, Michelle, we observed the 1999 Leonids from a cruise ship sailing near Samos, the island in the Aegean Sea where Pythagoras was born and the home of Aristarchus, who postulated a Sun-centered universe more than 2,200 years ago. If the island's sky was clear for Aristarchus, it certainly wasn't for us. The night of November 17th began with clouds and, other than a few periods of clearing, it stayed that way throughout. We did catch glimpses of several fireballs through holes in the cloud cover.

A Leonid meteor streaks past the Hyades over the famous Matterhorn. German amateur Roland Eberle captured this scene on November 17, 1998, from Gornergrat, Switzerland.

One of those holes was fortuitous, however. At around 4:10 a.m. local time, as the Leonids were coming down thick and fast over the Mediterranean, the clouds broke in one small window that revealed the stars of Auriga. We saw five bright meteors race through that small piece of clear sky in a single minute! So despite the less-than-ideal conditions, we can say that we did indeed see the Leonid storm of 1999.

The Earth had a rough year in 1999, sweeping up lots of meteoroids on several occasions. Less than a month after it had left the Leonid stream, the Earth encountered the regal Geminids. Unlike the Leonids, which plunge headlong into the upper atmosphere at 71 kilometers per second, the Geminids arrive at only about half that speed. There are few things in the sky more exquisite than a slow Geminid fireball gliding gracefully across the heavens, and we looked forward to seeing a few last December.

We were not disappointed. For two nights Wendee and I, instead of traveling halfway around the world, walked a few yards to our backyard observatory and watched the show, this time with astronomy entrepreneurs Bob and Lisa Summerfield. On the night of the 12th, the eve of the predicted maximum activity, we saw 309 Geminids. The next night was very special: we counted 750 over 6½ hours, including a faint sporadic meteor that climbed the sky slowly from the southern horizon and didn't disappear until it reached the zenith. The Geminids of 1999 closed the year on a high note by demonstrating that a good meteor shower can occur at almost any time.

I'll never forget my first tiny meteor that fell almost a half century ago. Nor will many readers forget that their first attraction to the night sky was sparked by the sudden appearance of a meteor that caused them to take a quick glance skyward, precipitating a lasting interest in astronomy. Imagine the journey of a Geminid meteoroid, drifting away from the asteroid Phaethon 100,000 years ago and orbiting the Sun for long ages. Imagine it making several close passes to Earth over several decades. Visualize the tiny particle closing in on Earth, watching the planet grow from a brilliant point of light to a monstrous blue-and-white marble filling half its sky. As the meteoroid zips into Earth's upper atmosphere, it heats the surrounding air to incandescence and vanishes forever.

Imagine finally that at the right moment a youngster looks up, sees the meteor, asks his parents, "What was that?" and starts, as many of us did, on a lifelong fascination with the sky.          APRIL 2000

# A Perfect Storm

At precisely 1:30 on the morning of November 19th (November 18th Universal Time), a Leonid fireball began as a bright bluish blob of light low in the northeast. Although the meteor shower's radiant in the Sickle of Leo was still below the horizon at that time, the fireball streaked high across the sky, moving slowly at first, then picking up speed as it arched right over our heads, sparkling silently like a fireworks display. Lasting more than 10 seconds, the fireball finally disappeared over the southwestern horizon, burning out after grazing Earth's upper atmosphere. "Everyone had time to look up," my wife, Wendee, noted, "spot the meteor, and give the appropriate shouts of approval. We knew right away we were in for a good show."

It was the single most dramatic meteor event I had ever seen, but on this night the fireball was only the opening act to our 2001 Leonid adventure. Our observing site was the Ewaninga Rock Carvings Conservation Reserve, a 6-hectare (15-acre) park featuring a dry lakebed and sandstone outcrops about 40 kilometers (25 miles) south of Alice Springs, in Australia's Northern Territory. When we stepped off our bus we immediately realized what a treasure this site was. In full view of an ancient Aboriginal rock carving, or petroglyph, we were under one of the darkest skies we had ever experienced.

Our tour was organized by Edith and Fred Stanton of Intrepid Traveler. Earlier that afternoon Edith had arranged a four-hour visit to a complex of impact structures known as the Henbury craters. We were concerned about the region's unstable weather that day — in fact, just a few hours before our Henbury trip it was raining in Alice Springs. But as the Earth began its 2001 encounter with the Leonids (it was already night over western Europe) the sky cleared up and we headed to the park.

When we reached Ewaninga's lakebed, we set up camp close to the rock carvings. As we began our Leonid vigil, we wondered what the Aboriginals, centuries ago, would have thought of the magnificent sky overhead — what legends and stories they might have developed.

It didn't take long after the first fireball before a second one lit up the southern sky, grazing the atmosphere like its predecessor. Meteors started to appear steadily after that, but as Earth approached the outer fringes of the debris trail that the Leonids' parent comet, Tempel-Tuttle, had shed during its return in 1699, rates increased rapidly. Surprisingly, the 1699 meteors were mostly bright, at least lst magnitude. For a short time around 2:45 a.m. local time, the 1699 debris trail produced a steady stream of meteors every few seconds.

The rates didn't subside after our encounter with the 1699 trail, but it was easy to see when Earth plowed into the comet's 1866 debris (the material that was released during the time when the comet was discovered independently by Wilhelm Tempel and Horace Tuttle). As Leo rose higher in the sky, suddenly, at around 3:15 a.m., a swarm of faint meteors with short streaks appeared near the radiant. It was as though someone were emptying a bowl of water out of Leo's Sickle — meteors were pouring out! At one point Wendee and I recorded nine meteors simultaneously, and there were many instances of five or more meteors coming in at the same time.

Meanwhile, the Southern Cross had risen in the southeast, dragging Beta and Alpha Centauri with it. As I was pointing out the magnificent expanse of the southern Milky Way to our group, two fireballs cruised right through the Cross. Eventually the full majesty of the southern sky unfolded, with Alpha and Beta Centauri, the Southern Cross, the Jewel Box and the Coalsack, the Eta Carinae Nebula, and the Magellanic Clouds all forming an exquisite backdrop to a storm of meteors.

So many meteors were falling that from our site some of the long-trailed ones seemed to converge toward the southwest opposite Leo as if forming an "antiradiant." I saw a Leonid appear to intersect with a sporadic meteor. Could this be what astronomical artist Leopold Trouvelot, who witnessed the 1868 Leonid display over Massachusetts, really saw when he painted a meteor that seemed to stop and do a sharp turn?

The zodiacal light brightened, then dawn began, and still the meteor rates kept up. They were still raining down well into bright twilight, when we saw our final meteor just 25 minutes before sunrise. I had always wanted to see a meteor storm; after November 19th's total count of 2,164 meteors, I finally did!                                         APRIL 2002

# Red Lights in the Sky

On the morning of November 23, 1984, at around 5:15 a.m., I began my comet hunt by systematically sweeping the southeastern sky. At that moment I was in a state of tremendous excitement — the last time I had searched this part of the sky with my 16-inch telescope, I actually codiscovered my first new comet.

Almost immediately after I began my sweep, I stopped dead in my tracks. In the telescope's field of view was a star in Hydra that was so red it looked like a traffic-signal light. I thought the object had to be a rare carbon star, one with an atmosphere rich in carbon compounds, which could account for its deep red hue. (Such a sooty atmosphere acts like a red filter, almost completely absorbing the blue and violet portions of the star's spectrum.) I later found out the star's true identity: V Hydrae.

The winter sky has two really famous red stars. The brighter one, Mu ($\mu$) Cephei, also known as Herschel's "Garnet Star," is roughly midway between Alpha ($\alpha$) and Delta ($\delta$) Cephei. Mu's spectral type is M2, which shouldn't result in its being that red, but it is. The other star is R Leporis, or "Hind's Crimson Star," probably the reddest Mira-type carbon star in the sky. First noticed by English amateur John Russell Hind in 1845, the star's brightness varies over a roughly 14-month period, from an average maximum magnitude of 6.8 to an average minimum of 9.6. And at spectral type C7,6e it's so red that visual observers have described it as like a drop of blood. (The "e" indicates the presence of emission lines in its spectrum.)

Red stars represent the late-type stars in the spectral classification system developed nearly 80 years ago at Harvard College Observatory. Originally, the system classified stars alphabetically, but astronomer Annie Jump Cannon, who studied the spectra of more than 325,000 individual stars, rearranged and merged the classes into seven basic groups: O, B, A, F, G, K, and M. This scheme is based on the stars' surface temperature — from the hottest deep-blue O stars to cool, red M suns. Subdividing each letter with numbers from 0 to 9, Cannon cut the sequence even more finely. For example, our Sun is a pale yellow G2-type star, Aldebaran is an orange K5, and Betelgeuse is a red M2.

The spectra of carbon (C-type) stars are similar to K- or M-type stars, but they also exhibit the strong spectral signatures of the carbon compounds $C_2$, CN, and CH. The variable star TX Piscium (19 Piscium), which varies from magnitude 4.8 to 5.9 every 220 days or so, is a very red C7,2. It's the brightest carbon star in the sky.

I recently discussed my accidental "discovery" of red stars with Elizabeth O. Waagen of the American Association of Variable Star Observers (AAVSO). She had a similar experience in the fall of 1977 while taking an astronomy course at Smith College, two years before joining the AAVSO staff. "I was doing a semester project on Cassiopeia for my introductory astronomy class," she remembers. "I was just browsing through the constellation with the college's 6-inch Clark refractor when I saw this beautiful star glowing ruby red." WW Cassiopeiae, it turns out, is a C5,5 star. According to the GCVS it varies irregularly from magnitude 9.1 to 11.7. When the star is this red, you don't find it; it finds you.

Red stars are fascinating to watch. Among the brightest of such stars in the Northern Hemisphere's winter sky is Mu (μ) Cephei, also called the "Garnet Star" by English astronomer William Herschel. It varies irregularly in brightness from magnitude 3.7 to about 5.0 over an average period of 730 days. Although it appears yellowish orange in this photograph, many visual observers report it as exhibiting a much deeper red hue.

I had that experience again last August 15th, when yet another red star, TU Geminorum, entered my telescope's field. This semiregular variable is of type C6,4, though it's not as red as V Hydrae (a C6,3e–C7,5e). TU Geminorum stays between magnitudes 7.2 and 8.5, and the GCVS lists its average period as roughly 230 days. Auriga has two nice red stars, both semiregular variables. S Aurigae, a C4–5,4–5-type star, ranges from magnitude 8.2 to 14.0 over a period of approximately 590 days. UU Aurigae, a C5,3–C7,4 star, stays fairly bright, varying from magnitude 5.3 to 6.7 in about 234 days.

What's the reddest star as seen through a telescope? That question is difficult to answer, for observers' eyes differ in their ability to determine star colors, especially red hues. You might be able to detect the degree of redness in a star more easily than someone else, and the field of view in which a star is located makes a difference; I find that red stars in rich fields, surrounded by a host of stars of other colors, look redder than similar stars do in sparsely populated fields. You also have to take into account the effects of atmospheric refraction, the darkness of the sky, the phase of the Moon (whose light makes the background sky look bluer), and the telescope's optics. In my opinion, the two reddest stars are V Hydrae and R Leporis.

The AAVSO has a massive series of star charts on which each variable is accompanied by comparison stars to help observers estimate its brightness. A word of caution, however: Very red stars are tricky to observe because red light tends to build up in intensity on your retina, like light accumulating on a photographic film. In what is known as the *Purkinje effect,* these stars will appear to brighten before your eyes if you stare at them. Use quick glances only when estimating a star's brightness.

About an hour or so after I observed V Hydrae in 1984, the night was interrupted by the rising of our nearest star. Right now, the Sun is a peaceful "main-sequence" star, but after some 5 billion years it will begin to change radically. At some point it will become a red giant, and then it could turn into a carbon star just before it sheds its outer layers to form a planetary nebula. Our Sun may not be massive enough to form such a nebula, but if it does it would be cosmically short-lived, lasting only 25,000 years or so. The thought that our own Sun, bright and warm as it is, could somehow have a future linked to those beautiful, rare red stars makes them all the more interesting and fun to observe. DECEMBER 2002

# A Tale of Two Eclipses

*Thy shadow, Earth, from Pole to Central Sea,*
*Now steals along upon the Moon's meek shine*
*In even monochrome and curving line*
*Of imperturbable serenity.*

*How shall I link such sun-cast symmetry*
*With the torn troubled form I know as thine,*
*That profile, placid as a brow divine,*
*With continents of moil and misery?*

*And can immense Mortality but throw*
*So small a shade, and Heaven's high human scheme*
*Be hemmed within the coasts yon arc implies?*

*Is such the stellar gauge of earthly show,*
*Nation at war with nation, brains that teem,*
*Heroes, and women fairer than the skies?*

— Thomas Hardy (1840–1928), "At a Lunar Eclipse"

"What's wrong with the Moon?" my five-year-old granddaughter, Summer, asked her parents in Albuquerque, New Mexico, last November 8th. "It looks spooky." Clear skies throughout much of the Americas and parts of Europe and Africa allowed millions of people to enjoy the beautiful total eclipse of the Moon that night. Across the northeastern United States, people braved subfreezing temperatures to stand outdoors on driveways and sidewalks, staring and pointing at the Moon.

I love the penumbral phases of a lunar eclipse — they allow unparalleled views of the long rays emanating from the craters Tycho and Copernicus, without the usual glare of a full Moon. But this particular penumbral phase was lighter and harder to detect than most. I wasn't sure I was seeing penumbral shading until about 20 minutes before the first umbral contact. Then, in the last five minutes before the partial eclipse began, the shadow's leading edge darkened rapidly.

Thomas Hardy was right: there is something magical about the onset of the umbra. It really "steals along upon the Moon's meek shine," and it really does give an impression of unbelievable peace. A century after the great English novelist wrote the sonnet above, our November 8th eclipse was gorgeous, although quite a bit brighter than usual, thanks to the Earth's atmosphere being relatively free of dust and haze. What all of us shared that evening was the spectacle of seeing the Moon pass through Earth's shadow as it orbits our planet—a stark reminder that we live in a world in space, that we're a part of a planetary system in motion.

One hundred years ago another lunar eclipse stirred watchers in a different hemisphere. The night was April 12, 1903, and Thomas Hardy watched the Earth's shadow cross over the Moon. I actually saw Hardy's eclipse—though not in the same year—and I saw it better than he did. Here's how.

The eclipse of 1903 was not total, but its maximum magnitude of 97 percent made it very nearly so. In fact, according to NASA astronomer and eclipse expert Fred Espenak, it was the longest partial eclipse of Saros 130. *Saros* refers to the 18.03-year period when a lunar or solar eclipse appears to repeat itself (with a westward shift in longitude where it is visible). Eclipses belonging to a particular saros series or family have very

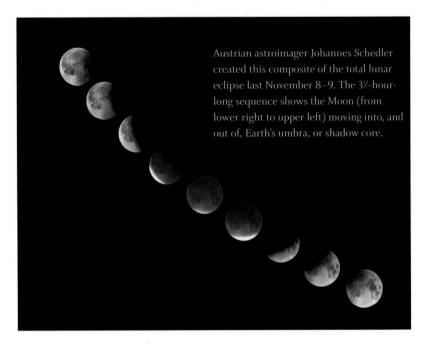

Austrian astroimager Johannes Schedler created this composite of the total lunar eclipse last November 8–9. The 3½-hour-long sequence shows the Moon (from lower right to upper left) moving into, and out of, Earth's umbra, or shadow core.

similar orbital geometry of the Sun, Earth, and Moon.

I saw Hardy's eclipse repeating itself on the morning of May 25, 1975, this time as a total eclipse, and it was visible over North America. At that time my friend Gerald Cecil, now an astronomer at the University of North Carolina at Chapel Hill, pointed out Hardy's sonnet. Thinking of the social problems of his time, just as eclipse watchers last November might have thought of ours, Hardy's work contemplates the eclipse in a way we observers rarely see.

Leslie Peltier once said that true amateur astronomers are poets. Maybe one needs to be a poet to appreciate the unbelievable beauty and majesty of an eclipse of the Moon. MARCH 2004

# A Not-So-Transitory Success

What did Ernst Mach (1838–1916), the Austrian physicist and philosopher who is famous for the Mach number — the ratio of an object's speed to that of sound — have to do with transits of Venus? On the early morning of June 8, 2004, I learned about Mach's unusual relationship with these rare astronomical phenomena simply due to the diverse background of the group I had observed with. We were on the summit of Westmount in Montreal, the same spot from which I saw my first eclipse (a partial solar one) in 1959.

My brother-in-law, Larry Stein, president of the Canadian Association of Radiologists, and Tim Hunter of the University Medical Center's Department of Radiology in Tucson, Arizona, immediately grasped the connection when, as Venus approached the Sun's southwestern limb at third contact, a hazy, dark line began to join the leading edge of the planet's disk with the solar limb. In astronomy, we know this as the "black-drop" effect. But in radiology, it's called the "Mach band," a form of optical illusion or visual edge enhancement produced by the normal physiologic processes of the human eye. On a chest X-ray, Larry and Tim explained, the heart appears much brighter than the lungs. Where the boundary of the heart is seen against a lung, it frequently appears to be edged in black. The thin black line between the bright heart and the darker lung (or, in this case, between the bright Sun and dark Venus) is a psychovisual effect. The rest of our group — Carol Hunter; Peter and Dianne Jedicke; and my wife, Wendee, and I — were so caught up in the discussion that morning that we almost forgot the real reason we'd traveled to Montreal: to witness the first transit of Venus in nearly 122 years.

We arrived at this bustling city in southern Quebec a few days early to allow for extra time in case we needed to drive far in search of clear skies. Montreal is Canada's second-largest city, and even though it's situated on an island in the St. Lawrence River, its extensive road system makes it easy to leave quickly and head off in any direction. In the week leading up to the transit, however, the daily weather forecasts promised a cloud-free sky on transit day. That's why we were a bit surprised to see a deck of clouds cover the sky as night fell on the eve of the event. Still, a careful study of the latest weather-satellite images indicated that the sky would clear up by midnight. With that happy thought in mind, we went to sleep for a few hours.

I awoke at 4 a.m. to find the sky completely overcast. This time the

Last June 8th the Sun rose over the eastern coast of North America with the transit of Venus already in progress. Jeff Ball snapped this view over the Atlantic Ocean from South Carolina's Huntington Beach State Park.

satellite pictures indicated a clearing some 100 miles to the west. We decided to assemble at Westmount's lookout and plan what to do next. With dawn advancing rapidly, we somewhat morosely prepared to head to Ottawa. But just then there appeared a series of very small patches of blue sky. I turned to the west, where the clearing was supposed to be coming from, and saw a much larger clear patch along the horizon. We decided to stay put. Within 15 minutes the sky was almost completely clear, and as the Sun slowly rose above the clouds we saw what Wendee described as a hole punched through the Sun — the transit was in progress!

Over the next two hours, we were filled with two complementary emotions. The first one was simple: as Venus slowly made its way across the Sun's face, we marveled at the magnificent clockwork precision of celestial mechanics. The second emotion was more complex: we felt as though we were standing with every group that had ever seen, or had failed to see, a transit of Venus in the past four centuries.

With less than five minutes to go till third contact, scheduled to occur at 7:05 a.m. local time, we started to notice the thin, dark, hazy layer — the Mach band or black-drop effect — between the limbs of Venus and the Sun. It was then that the biggest miracle of all took place. Through the white-light filter on my 3.5-inch Questar telescope, I could clearly see the outline of the portion of Venus that had already left the Sun, and for about a minute after fourth contact, at 7:25 a.m., I could still see the outline of the entire planet. For me to see such a sight, the atmosphere of Venus must have been almost as bright as the Sun itself.

So ended our view of one of nature's rarest and grandest spectacles. It turned out that our choice of viewing site was a success in every way. We saw the transit under clear skies and almost perfect seeing. As the Sun continued to rise, clouds began to form again, but we didn't care; the transit that we had all waited a lifetime for was over, and we had seen it!                                                  NOVEMBER 2004

### Author Update:

*The next transit of Venus occurs on June 5–6, 2012. Wendee and I plan to watch it from our home in Arizona.*

# Seeing Einstein's Gravity Lens

On a bright afternoon in September 1979 my astronomer friend Gerald Cecil and I joined Dan Brocious, the public-affairs officer for the Smithsonian Astrophysical Observatory, on a trip up the long, narrow, winding dirt road to the summit of Mount Hopkins near Amado, Arizona. There Brocious gave us a tour of the new Multiple Mirror Telescope (MMT), an innovative instrument consisting of six identical 1.8-meter (72-inch) mirrors working in unison on an altazimuth mount. He also showed us a sample of a tiny new CCD chip, one of the earliest ever used for astronomical imaging. He then pointed out that these new technologies had already contributed to an amazing achievement: the discovery of the true nature of Q0957+561 A and B, the 17th-magnitude "twin" quasar in Ursa Major.

Also referred to as the double quasar, the twin quasar was discovered by astronomers Dennis Walsh, Robert F. Carswell, and Ray J. Weymann in late March that year with Kitt Peak's 2.1-meter reflector. They found the quasar's components to have nearly identical spectra and redshifts. The following month the team conducted follow-up observations, this time using the MMT, to verify its findings, which suggest that a massive, very dim elliptical galaxy lies in our line of sight to the twin quasar. (The light that we see now left the quasar 9 billion years ago.)

Subsequent observations from Palomar and Mauna Kea finally revealed the galaxy, which is situated almost directly in front of B, the quasar's southern component. This confirmed the original assumption of Walsh and his colleagues—that the quasar is a single object, and that the intervening galaxy, as well as the surrounding cluster of galaxies, acts as a gravitational lens, splitting the distant quasar into two separate images.

This notion that a massive body could bend the light from a more distant object was predicted by Albert Einstein in his general theory of relativity, which he published in 1916. The discovery of the twin quasar was the most significant find since English physicist Arthur Eddington recorded the gravitational bending of light from the Hyades star cluster by the Sun during the total solar eclipse of 1919.

Although astronomers have identified close to 100 examples of gravitational lenses in the universe, including the exotic Einstein Cross in Pegasus, I've always had a special feeling for the twin quasar, and recently I decided to try to resolve it visually with my backyard telescope.

This feat had been done before. On June 6, 1988, Brent Archinal (then with the US Naval Observatory) and amateur astonomer Bob Bunge saw Q0957+561 A and B with the Richland Astronomical Society's 31-inch f/7 reflector in Mansfield, Ohio. *S&T* contributing editor and renowned visual observer Stephen James O'Meara also reported splitting the quasar in 1991 with a 20-inch Dobsonian at the Winter Star Party and with a 25-inch reflector at the Texas Star Party.

On Valentine's eve this year, comet discoverer Rolf Meier and I, with our wives, Linda and Wendee, attempted to detect the quasar visually. Since the 11th-magnitude edge-on spiral galaxy NGC 3079 points almost directly to the quasar, just 10 arcminutes to its north-northwest, we all had an easy time locating the quasar's field. But spotting the quasar itself was a challenge — in my 16-inch reflector at 156× the quasar appeared as a very dim star.

Splitting the quasar's components, only 6 arcseconds apart, was another story. Heading over to Tom Glinos's 25-inch Ritchey-Chrétien tele-

*Below left:* Tom Glinos photographed this 21′-wide field including Q0957+561 A and B (arrowed) in Ursa Major last March 20th using a 25-inch f/8 RC Optical Systems Ritchey-Chrétien telescope. The object is the first gravitationally lensed quasar ever discovered. It is located 8½° west of Merak, Beta (β) Ursae Majoris, at right ascension 10ʰ 01.4ᵐ, declination +55° 54′ (equinox 2000.0). *Below right:* Components A and B are 6 arcseconds apart. North is up, and east is to the left.

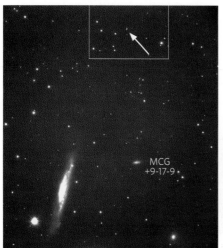

scope, we inserted a 7-millimeter eyepiece to obtain 726× magnification. When an object is near the limit of visibility, it doesn't appear constantly in the field but tends to pop in and out of view, depending on the seeing conditions. For me, during moments of steady seeing, the quasar appeared as two closely spaced but clearly separate objects. Three weeks later, on March 9th, a night with exquisite seeing, I saw the quasar's components again, this time through my 16-inch at 581×.

The twin quasar is difficult to detect, but it can be done with large apertures and very steady skies. Its proximity to NGC 3079 is a big help in locating it. After you find its general area, use averted vision to spot the quasar at the end of the quadrilateral asterism of stars. Be patient; it isn't an easy target. The quasar should appear slightly elongated. Finally, if the seeing is particularly good, try to resolve its components. If you succeed, you'll see gravitational lensing — and Einstein's idea — in action.

JULY 2005

# In Praise of Penumbral Eclipses

During my long voyage to the South Pacific to view the April 8th total solar eclipse — which I wrote about in the August Star Trails (page 107) — I teasingly asked my fellow eclipse chasers where they were planning to go for the penumbral eclipse of the Moon that would follow two weeks later, on April 24th. Their expressions were universally incredulous — why would anyone observe such a thing, let alone travel to see one?

I *love* eclipses. And that passion includes *all* events involving the alignment of the Sun, Earth, and Moon — from partials and penumbrals to annulars, totals, and annular-totals. But many may ask, why bother to observe those events where the Moon fails even to enter Earth's dark umbral shadow? The best you can see during a penumbral lunar eclipse is slight shading on the edge of the Moon closest to the umbra. Personally, I believe these events are great; it's fun to watch Earth's faint outer shadow glide slowly across the lunar limb, causing mountains and lava plains to appear in stark contrast to the full Moon's bright, nearly featureless disk. A penumbral eclipse also allows the rays of craters, such as Tycho and Copernicus, to stand out more prominently than at any other time of the lunar cycle.

So it was with excitement that I set out to catch the April 24th penumbral event amid threats of clouds and rain. The Clear Sky Clock (http://cleardarksky.com/csk) indicated the possibility of a hole in the cloud cover to the west of my home in Vail, Arizona. When my alarm clock went off at 2:25 a.m., I looked outside the bedroom window. The sky was still overcast, but at least it had stopped raining. At 2:40 a.m., with just 15 minutes to go till mideclipse, I decided to load Echo, my trusty 3½-inch Skyscope reflector, into the car and head west.

The farther west I drove, the more promising the sky appeared. This early in the morning, Echo and I had the road pretty much to ourselves. About 30 kilometers (20 miles) from home, I reached a deserted intersection marked by a lone traffic signal. I pulled over, took the scope out of the car, and waited a few minutes for the Moon to appear from behind a cloudbank. My plan was to handhold the telescope just in case I needed to make a quick getaway from this instant roadside observatory.

Even through thin haze, I could clearly see the darkening on the Moon's north-northwestern limb. I guessed that, had this been a total

Tenho Tuomi, produced this montage of the April 24th penumbral lunar eclipse from images taken with an 8-inch reflector and a digital camera. It shows the setting Moon as it skimmed through the penumbra from west (right) to east. "There was a thin cloud covering the Moon in my last shot," says Tuomi.

eclipse, it would have been a bright one — with a Danjon luminosity (L) rating of 3 or 4, typical of totalities wherein Earth's atmosphere is relatively dust-free. I was feeling so good, having beaten the odds to catch the eclipse just a little past its midpoint, that I barely noticed the approach of another car. I turned around to see a police cruiser pull up right behind me, its strobe lights flashing and its occupant nervously eyeing the long black tube I was holding.

I walked over and explained to the officer that I had a telescope in my hands and that I was using it to see the eclipse. Right away he looked at the Moon and noticed its darkened edge. After a brief conversation about eclipses, he bade me a good morning and left. With the onset of more clouds, I headed home, but not before I stopped by a beautiful grove of giant saguaro cacti, the same spot where I'd captured some pretty neat pictures of Comet Hyakutake nine years ago.

Finally I was home, my telescope safely set up in our backyard. It was now an hour past mideclipse, and a little less than half of the Moon still was within the penumbra. Surprisingly, even with its unusual lightness, I could still make out the shadow's presence on the Moon's northeastern limb. I had never detected a penumbral eclipse with that little of the shadow on the Moon before. (Conventional wisdom agrees that it's only when the Moon is halfway into the penumbra that the shading becomes barely discernible.) Ten minutes later, I could no longer detect any trace of the penumbra.

So ended my early-morning adventure. According to NASA eclipse expert Fred Espenak, this particular eclipse, which belongs to saros 141, was the 23rd of a series that began as an undetectable penumbral event on August 25, 1608, the year William Shakespeare registered his play *Antony and Cleopatra*. Before April 24th, the last penumbral event I saw belonging to saros 141 was on April 14, 1987. The eclipse will go deeper and deeper into Earth's shadow until it finally enters the umbra and produces its first partial eclipse, on May 16, 2041. Its first totality occurs on August 1, 2167. Then the saros series fades away, ending with another undetectable penumbral event on October 23, 2906.   SEPTEMBER 2005

# A Ringside Seat

"The annular eclipse of the Sun last October 3rd," says my wife, Wendee, "had its own special magic. It wasn't the gut-wrenching splendor of a total solar eclipse; it was a quiet wonder as the uneven circle of Baily's Beads began to form and then turn into a beautifully symmetrical ring."

This was the fourth solar event I've seen that had taken place in October. The first one was a partial eclipse on October 2, 1959, from Montreal. The second was a very shallow partial eclipse on October 3, 1986, from the ghost town of Steins, New Mexico. Eighteen years, or one saros cycle, later, the partial eclipse was repeated, this time on October 13, 2004, from Mauna Kea. The 2005 annular eclipse in Madrid, Spain, was one Metonic cycle from the 1986 eclipse. Under this cycle, it's possible for a series of four or five eclipses to occur on the same dates but 19 years apart.

We traveled to Spain and met a group of about 20 eclipse enthusiasts from Switzerland, Sweden, Denmark, and the US. As the first rays of the Sun hit Madrid's splendid royal palace on eclipse morning, the whole city seemed bright with anticipation. Not since April 1, 1764 — more than 241 years ago — had the city experienced an annular eclipse. My guide, Fernando, and I drove to Madrid Park, where we saw dozens of telescopes being set up along the edges of a large pond. The excitement was intense, a further testament to the single language that lovers of the sky share, especially when Mother Nature is about to put on one of her grandest shows.

We headed back to our observing site, on the roof of our hotel. Once there, Wendee and I set up Minerva, our 6-inch reflector fitted with an improvised solar-filter mounting made out of a flowerpot holder. Piggybacked to Minerva was a Coronado PST hydrogen-alpha (Hα) telescope. As the Moon took its first

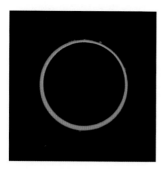

Wendee Levy snapped this view of the October 3rd annular eclipse from a hotel rooftop in Madrid, Spain, using a Coronado PST hydrogen-alpha telescope and a digital camera.

bite of the Sun's western limb at 9:40 a.m. local time, the spotless Sun sported a magnificent cluster of small prominences on its opposite side. As the eclipse deepened, the whole city gradually dimmed. The temperature plummeted as a brisk wind picked up. A mile or so to the west, the royal palace also darkened slightly, just as it had in 1764. Office workers began filling the rooftops all over the city, all gazing skyward.

As I peered through Minerva's eyepiece, the scope's white-light filter allowed me to see a thin, horseshoe-shaped patch of light along the Sun's limb. But looking through the Coronado PST, Wendee suddenly called out, "There's a small red dot appearing at the very open end of the horseshoe!" It was a mighty prominence just coming into view. Right at that moment, we heard an announcement from somewhere urging everyone to go indoors to avoid the "harmful" solar rays.

Two minutes later, a perfect ring of sunlight blazed in the clear blue skies above Madrid. As the Moon slowly marched across the Sun's face in the next four minutes, through the Hα filter it really looked like a total eclipse, with the darkened center of the Sun surrounded by prominences. The only thing missing was the corona. Looking around from our rooftop perch, we could see that Madrid was virtually at a standstill, as people everywhere stopped what they were doing, donned their solar glasses, and looked to the sky. After annularity, the partial phases took over once again. We saw two "last" contacts, the white-light one taking place about 30 seconds before the one seen in Hα.

After eclipse day we journied to Barcelona. Here, almost exactly 400 years ago, inhabitants witnessed the great European total solar eclipse of October 12, 1605. It was the same eclipse that Shakespeare wrote about in *King Lear* and the one that prompted Johannes Kepler to suggest that the corona belonged to the Moon. The eclipse track passed through Barcelona, immersing the city in darkness. Our next chase is in the Mediterranean Sea for the March 29, 2006, total solar eclipse. See you there!

FEBRUARY 2006

## Author Update:

*The annular in Spain was my 70th eclipse, a tally that ranges from barely detectable penumbral lunar eclipses to total solar and lunar eclipses. In 2006 I added two more to my total.*

# My First Meteor

Fifty years ago this summer — on July 4, 1956 — a meteor sparked my interest in the night sky. As it flared in the upper atmosphere, it planted a seed in my young mind.

At the time I was at Twin Lake Camp, located off Route 30 in central Vermont and less than a two-hour drive from Stellafane. This was my first summer away from home, and we were having an Independence Day celebration in the camp's social hall. When it ended, our group of 8-year-olds made its way back toward Bunk B. I happened to look up and saw the meteor streak across a small portion of sky. It wasn't particularly bright — perhaps 2nd magnitude — moving fairly slowly and leaving a short trail that headed down toward a bright star that was almost certainly Vega. It lasted only about a second. I asked my friends if they had seen it, and when they told me they didn't, I decided that the shooting star was meant for me.

It took a while for something to sprout from that seed. Fifteen months later, I heard about the launch of Sputnik 1 in October 1957. Two years after that, the solar eclipse of October 1959 attracted my attention. Finally, at the start of the summer of 1960, a cousin gave me a book about planets as a get-well present for a broken arm I suffered in a bicycle accident. By the end of that summer, the seed had matured into a lifelong passion.

So many memories of youth get lost over the years, but not the me-

I'm the second boy from right in this shot at Twin Lake Camp in the summer of 1956. My sighting of a meteor, possibly an Omicron Draconid, on the Fourth of July that year led to a lifelong fascination with the stars.

teor incident. I recall precisely where I was at Twin Lake when I saw it. But initially what I couldn't remember was the exact year.

I spent three summers at Twin Lake — in 1956, 1957, and 1958. For a long time I had assumed my meteor sighting happened during my final year when, at age 10, that experience would have had more meaning. But two years ago I did some research into old weather reports. Stellafane's John Martin provided me with Vermont newspaper records indicating that the sky on Independence Day 1958 was rather stormy. Further communication with the National Weather Service showed that the sky probably wasn't particularly clear that night in 1957 either, as clouds and isolated thunderstorms rumbled through the state. Besides, any planetarium program would reveal that there was a Moon in the sky that evening. My memory is of the clearest, darkest, and most star-filled sky I had ever seen. Weather records indicate that the night of July 4, 1956, was clear throughout Vermont. So the magical night must have taken place that year, especially given that the nearly new Moon was absent from the sky at the time!

Now that I had the date confirmed, I wanted to learn more about the meteor. I went back to the camp and stood at the same spot, trying to reconstruct the meteor's path toward Vega. There are two minor streams that begin their activity around that time. One, the Alpha Lyrids, lasts from July 9 to 20, but it features mostly faint telescopic meteors. Its naked-eye rate is about 1 to 2 per hour. Besides, my meteor sped *toward* Vega, not away from it.

The other shower is the Omicron Draconids. My meteor was a night short of the shower's traditional opening night of July 5–6, but the date is close enough. This shower was discovered by a team of astronomers, including Allan Cook and Brian Marsden, who noted three meteors on Harvard College Observatory photographs taken between 1952 and 1954. The team proposed in 1973 that there was weak evidence linking the shower to comet C/1919 Q2, which was discovered by Unitarian minister Joel Metcalf.

If Metcalf's comet is the actual parent of the Omicron Draconids, then it's a fortunate Vermont coincidence. Metcalf found the object from another Vermont camp northwest of Twin Lake in the summer of 1919, the same year Twin Lake first opened. While I cannot be certain that my half-century-old meteor sighting was actually an Omicron Draconid, I suspect that it was, and it's whimsically possible that my gaze skyward, at age 8, is the first visual observation of an Omicron Draconid.

It takes so little to begin a lifelong interest in the stars, and it's best if the sky gets the chance to light the fire all by itself. How many of us got started by watching an occultation or an eclipse? These things demonstrate that the celestial tapestry above us is very dynamic and constantly changing, and that a chance encounter with a meteor from a long-gone summer can change the course of a life. OCTOBER 2006

# What Is a Planet?

The question of how to define a planet has been a deeply personal one for me since a summer evening in 1960. As I listened to Dad's recounting of the story of how Uranus, Neptune, and Pluto were discovered, I sat on the edge of my chair at the dinner table. To me at that time, the search for planets was a tale of high adventure. How many planets, I asked, are there in our solar system? "Nine," Dad answered.

The Sun's family was much simpler back then. Now we know that Clyde Tombaugh's discovery of Pluto in 1930 is far more complex than was imagined. Though it took more than six decades, Pluto opened the door to a whole region of the outer solar system we now call the Kuiper Belt. We already have Eris (formerly Xena or 2003 UB$_{313}$), Sedna, and

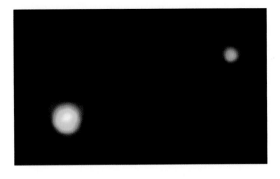

In 1994 the Hubble Space Telescope gave the world its first view of Pluto (left) and its moon Charon as separate, sharply defined round worlds. The two were about 20,000 kilometers (12,000 miles) apart and 4.4 billion km from Earth at the time.

Quaoar, and we'll surely find many more Pluto-size wanderers.

For me, Dad's story picks up again with the International Astronomical Union's planet-definition committee, which attempted to put the solar system into perspective last August, during the IAU General Assembly in Prague, Czech Republic. The members of this committee, chaired by Owen Gingerich, a highly regarded Harvard astrophysicist and historian of science who is familiar with how humanity's understanding of the solar system has evolved over the years, should be congratulated for their Herculean effort in coming up with a formal definition based entirely on gravity. Such a definition may not seem terribly important to astronomical research, but it's vital to the public's perception of astronomy, especially for children trying to make sense of their environment.

Unfortunately, what I saw happen both at the IAU meeting and among other groups of astronomers was that many of them made their decisions long ago. Most debaters in Prague were talking past each other like worlds passing by in separate orbits.

Dad's Pluto story in 1960 didn't include its status as a planet, even though British astronomer Raymond A. Lyttleton suggested around 1936 that Pluto could be an escaped satellite of Neptune. In 1980, during a ceremony honoring the 50th anniversary of Pluto's discovery, a proposal was made to "downgrade" it to an asteroid, and the debate has been raging ever since, just as the Pluto system has been getting progressively more interesting as we learn more about it.

This adventure story of discovery and learning is so much richer than what the news media has been advertising lately. For many of them, it seems the only headline was that on August 24, 2006, the IAU demoted Pluto. If you read all the articles and editorials that have poured into newspapers and Web sites all over the world, that is primarily the story you get. But if you read all the parts of the planet definition that was officially adopted by the IAU, there's more to it than that.

We now have planets and dwarf planets, the only difference being seven words that say that a planet "has cleared the neighborhood around its orbit." It can be argued that a "dwarf planet" is a form of planet (just as our Sun, a dwarf star, is a star), regardless of what the intent of the IAU might have been. So what happens when a child asks how many planets there are in our solar system? My answer: 8 planets, 3 dwarf planets, and countless small solar-system bodies.

"A dwarf planet is still a planet," notes James Christy, who discovered Pluto's moon Charon in 1978. "I'm surprised at the reaction that somehow there are now only eight planets. There is no way in the English language for an adjective to alter the basic meaning of the noun. The new terminology follows from stellar terminology where we have giant stars and dwarf stars. Dwarf stars are stars and are included in any counting of stars. Therefore, we now have 11 planets, with more to come."

A diverse group of professional and amateur astronomers, including Alan Stern of NASA's New Horizons mission now en route to Pluto, have launched an effort aimed at reaching a better definition.

"A small fraction of the IAU membership has forced the IAU into a technically and linguistically flawed definition of planethood," says Stern. "I believe the position the IAU is in now is untenable, and will therefore result in a general revision of their definition — if not by the IAU, then by the public and working research community."

Whatever happens, the process of discovery and understanding of our solar neighborhood continues. New objects will be found, new theories will be formed, and over time our picture of the Sun's family will be transformed as fresh data comes to light. But none of this should affect our desire to go out into our backyards and observe these objects.

For example, Ceres isn't just a binocular object; it is history. When I look at Ceres, I recall that at its discovery in 1801 it was called a planet, a status that was taken away about a half century later. I also remember the excitement generated by Christy's discovery of Charon, which increased immeasurably the complexity of the Pluto system, and how members of the New Horizons science team in 2005 discovered Nix and Hydra, two more moons of Pluto, even before their spacecraft left the launch pad. These are the adventure stories of astronomy, the history that makes us love what we do. Those seven committee members and the IAU during the summer of 2006 added a new chapter to that adventure. It won't be the last one.                                    DECEMBER 2006

### Author Update:

*My biography of Tombaugh (Clyde Tombaugh: Discoverer of Planet Pluto) is available from Sky Publishing. I thought it was nice of them not to change the title, even though Pluto is now only a "dwarf" planet.*

# Index

# Acknowledgements

Since my Star Trails column began in 1988, I have worked with many people in several departments at *Sky & Telescope* magazine. I especially want to thank Stephen James O'Meara, my editor from 1988 to 1995; Edwin Aguirre, my editor from 1995 to the end of 2006; and Paul Deans, who has done a remarkable job of turning my columns into a book. I'd also like to thank Leif Robinson and Rick Feinberg, *S&T*'s editors in chief during the run of my column.

# Image Credits

## Preface

**p7** David Levy

## Ideas

**p13** Gary Rosenbaum and Steward Observatory, the University of Arizona; **p15** Greg Pyros; **p17** Sky Publishing: Chuck Baker; **p20** Peltier Family; **p23** Wendee Wallach-Levy; **p26** William K. Hartmann (x2); **p29** [middle] Akira Fujii, [bottom] David Levy; **p31** Sky Publishing: Craig Michael Utter; **p35** E. C. Krupp; **p41** [left] David Levy, [right] NASA/JPL/University of Maryland; **p43** Mark Vigil.

## People

**p50** Fred Espenak; **p53** Bob Summerfield; **p54** Wendee Wallach-Levy; **p57** [left] Sky Publishing: J. Kelly Beatty, [right] NASA/Carolyn Porco; **p61** Yerkes Observatory; **p65** Wendee Wallach-Levy; **p67** Montreal Centre, RASC; **p70** NOAO; **p73** Scott Tucker; **p77** [left] Akira Otawara: Temmon Guide, [right] Shigemi Numazawa; **p81** Tsutomu Seki.

## Places

**p85** Steve Buhman, Southern Illinois University Carbondale; **p89** Barry Simon; **p92** Wendee Wallach-Levy and Kitt Peak; **p94** Sky Publishing: J. Kelly Beatty; **p97** NASA; **p99** Haldun I. Menali;

## Things

**p103** Akira Fujii; **p107** Sky Publishing: J. Kelly Beatty; **p110** David Levy; **p112** NASA; **p116** David Levy; **p119** Mark Vigil.

## Objects & Events

**p123** Regal Empress courtesy Joe Rao; **p126** Roland Eberle; **p131** Akira Fujii; **p134** Johannes Schedler; **p136** Jeff Ball; **p139** Tom Glinos (x2); **p141** Tenho Tuomi; **p143** Wendee Wallach-Levy; **p145** David Levy; **p147** NASA/HST.